WINNING THE RADAR WAR

WINNING THE RADAR WAR

A Memoir

by Jack Nissen
and A.W. Cockerill

St. Martin's Press
New York

All photographs are from the personal collection of Jack Nissen.

Library of Congress Cataloging-in-Publication Data

Nissen, Jack Maurice.
          Winning the radar war / byt Jack Nissen with A.W. Cockerill.
              p.   cm.
        ISBN 0-312-01535-6 : $19.95
          1. Nissen, Jack Maurice.   2. World War, 1939-1945—Radar.
3. World War, 1939-45—Personal narratives, English.   4. Great Britain.
Air ministry— Biography.   5. Engineers—Great Britain—Biography.
I. Cockerill, A. W.   II. Title.
D810.R33N57 1988
940.54′41—dc19   87-27123
                                                                       CIP

First published in Canada by Macmillan of Canada, a division of Canada Publishing Corporation.

First U.S. Edition

10 9 8 7 6 5 4 3 2 1

To
Sir Victor Hubert Tait KBE,
the quiet Canadian who
masterminded the radar war

CONTENTS

Maps   vi–vii
Introduction   ix
Chapter One Recruit to Secrecy   1
Chapter Two Pre-War British Radar   17
Chapter Three Pre-War German Radar   30
Chapter Four The Outbreak of War   40
Chapter Five A Scottish Testing-Ground   54
Chapter Six Deception and Interception   67
Chapter Seven Radar and the Battle of Britain   78
Chapter Eight The Tizard Mission to Ottawa
and Washington   94
Chapter Nine Ghost Station   102
Chapter Ten Night Attacks   116
Chapter Eleven Navigational Radar —
The Tide Turns   130
Chapter Twelve "Jubilee"   145
Chapter Thirteen Freya 28   162
Chapter Fourteen Dieppe Post-Mortem —
The Tide Builds   192
Chapter Fifteen D-Day   204
Afterword   211
Index   217

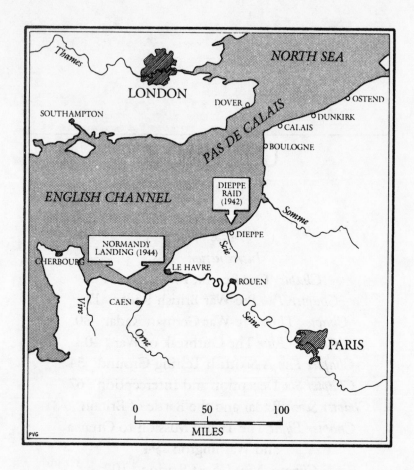

MAP 1   Northwest coast of France

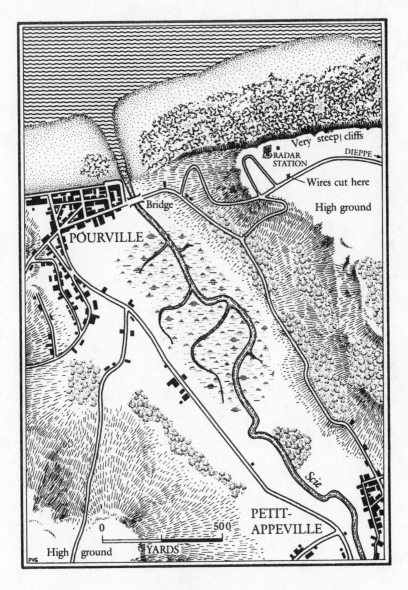

MAP 2   Area around Pourville and Freya 28
radar station

INTRODUCTION

"Secret operations are essential in war;
on them the army relies to make its every move."
Sun Tzu (circa 500 BC)
The Art of War

Secrecy is the handmaiden of war. She is the eyes and ears
of warriors in battle and lurks behind the lines when the
combatants retire to catch their breath. Without her help,
the strongest army, as a champion blinded, will falter. This
was true in the age of Sun Tzu, in the age of the classical
Hellenes and Romans, and in the age of medieval warriors
— and it is just as true of modern warfare. The need for
commanders of armies and opposing states to prepare in
secrecy, to keep secret their assets and resources for waging
war, to deceive the enemy about their intentions, is as essential today, with all the modern weaponry opposing states
have at their disposal, as it was in ancient times.

Writing in 1940 to explain the fundamental difference
between the First World War and the new great war that had
engulfed Europe once again, James H. Burnham held that
the First World War was the last war of capitalist states. The
conflict which by then had broken out, he said, was the first
war of the "Managed Societies". He meant that modern
nations were dominated and controlled by a new breed of
rulers whom he called "The Managers".

True to the tenets of their breed, the managers wrestled with and developed new production techniques, perfected the assembly lines invented by Henry Ford, and, by their industry, assured that the pipeline supplying armaments to their warrior companions was never empty; at least, it seemed never empty. As a result, the warring states were themselves nothing less than enormous machines of production that needed three things to keep them going: raw materials to feed them, workers to operate them, and "markets" in which to squander the output. In this, the United States of America alone had the organization, strength, and capacity to dwarf every other national economy. At the height of the war, for instance, the city of Pittsburgh, Pennsylvania, was producing more steel than the Axis powers combined.

The same warring states assembled massive armies and the most modern weapons that scientists could devise. Bomber fleets that the Axis powers had tested in the 1935–36 Italian invasion of Abyssinia (now Ethiopia) and the Spanish Civil War were expanded for the European conflict. Tanks, guns, warships, and aircraft were available in abundance. Yet secrecy was still the handmaiden. Without her, the mighty armies would dash themselves like waves on a rocky shore and vanish without a trace. All the modern weaponry would be wasted if not skilfully directed on the basis of secret intelligence.

The Axis powers had the advantage of physical strength, and of national economies directed to provide fodder for their war machine. They were, on the face of it, invincible. In contrast, the Allies had the Ultra secret, the Enigma encoding and decoding machine that revealed the enemy's secrets, a knowledge that was worth the strength of Hercules, though this was not at first obvious.

Ill-prepared for war, British Prime Minister Neville Chamberlain, for all the criticism levelled at him because of his September 29, 1938, Munich Agreement with Hitler, gained a breathing-space of a year or more in which to make ready for the onslaught. That was insufficient time. With the opening of hostilities, the French and British suffered crushing defeats that brought about the evacuation of Dunkirk. But despite the setbacks, the British at least, under the leadership of Churchill, maintained the will and the determination to wage war. Although they had little enough in the way of arms and equipment that would make the determination anything more than frustrated resolve, they did have a secret that proved in the end to be a passport to freedom for the entire nation.

What was it? Not the Enigma encoding-decoding machine that gave the British a ringside seat at the enemy war councils and made them privy to German radio-communication traffic — although Enigma was a major source of intelligence. Nor was it their vast network of intelligence agents — although the network was of inestimable value to Allied prosecution of the war.

The key factor, the major secret hitherto unacknowledged, was that the Allies won the "radar war". Allied successes at sea, on land, and in the air during the six long years of the conflict were, it is true, achieved by well-equipped men and women dedicated to the cause. But those efforts were made more effective by the British development and production of radar devices and systems. That development was central to winning the Battle of Britain, the U-boat war, the great bombing war over Germany, the invasion of Europe, and the Pacific War. No other single factor was as influential in the ultimate outcome of the entire war as the development of radar.

This book is a history of the radar war, a war unknown to the general public and to most of the fighting men and women who took part. It was an exhausting battle of intellectual titans involving successes and failures, unwarranted assumptions, and gross misconceptions on both sides. The book is a personal view of that war as witnessed by one of its participants, Jack Nissen.

Jack Nissen is one of the few men still living who was part

of the radar war from the beginning. He was a teenager when it began, a bright young graduate of the new technical schools in radio and television theory and a willing recruit to the secret community of RDF scientists.

RDF was the early name of the technology; the acronym radar, for 'radio detection and ranging', came into use about two years into the war.

For this reason this is as much Jack Nissen's story as the history of radar from its beginning around 1930 to the end of the war in 1945. It is a gripping story of intrigue, danger, courage, and valour. Although the subject is highly technical, the book has been written for the general reader, not the specialist. It must be added, however, that one cannot discuss radar without including some explanation of technical detail. The features that differentiate various devices would have little meaning if they were not explained to some degree. For this reason, both technical readers and non-technical readers will have to make some allowance for our compromise treatment of the subject.

A. W. Cockerill

WINNING THE
RADAR WAR

1

RECRUIT TO SECRECY

In the late summer of 1937, a few weeks short of my eighteenth birthday, I made my first trip to Bawdsey Manor on the east coast of Suffolk, the family home of Sir Cuthbert Quilter. Sir Cuthbert, musician and composer, is remembered today mostly for his revival of English folk music. There was a sense of remoteness and isolation about this part of the English countryside, which betrayed few signs of life: farm workers gathering the harvest, little or no traffic on the roads, and only the odd pedestrian in the hamlets through which we drove. In winter, I discovered, it was difficult to imagine a more desolate and forbidding stretch of coast, with its deserted, windswept shore and the harsh beat of waves breaking on the pebbly beaches.

The house with its nine turreted and crenellated towers was built from Suffolk grey stone. The sombre oak doors and casement windows helped increase the sense of mystery. The appearance of the grey house was vaguely familiar, a domesticated Victorian homage to the Tower of London. It stood isolated in pastureland dotted with clumps of elm, beech, and oak. Located a little way inland, near the mouth of the River Deben, which flows into the Thames Estuary, the mansion was approached along a tree-lined lane that climbed in a steep gradient from what could only be described as an eighteenth-century-quality road.

Looking back to that far-away summer's evening of my youth, I recall feeling excited and intrigued by the sense of mysterious expectation, yet I was unable to grasp the significance of its importance to my future. I knew only that someone in authority was interested in my knowledge of wireless (radio) and television engineering. Radio was in its infancy and television hardly yet born.

Two years earlier, in 1935, the Scottish scientist Robert Watson-Watt secretly demonstrated that the detection of aircraft was possible by the use of radio transmission. At about the same time, German scientists and radio engineers made a similar discovery. This remarkable coincidence led to the independent development of two systems of range- and direction-finding: Funkmess by the Germans, and RDF by the British. The results of both German and British research in the field have given the world the complex radio-telecommunication systems that have become part of modern society.

My own involvement in the development of radar, of which my visit to Bawdsey Manor in 1937 was a harbinger of things to come, had its roots in a childhood fascination with wireless technology. In 1929, at the age of ten, I built my first radio and was fortunate a couple of years later to attend the Mansford Technical Secondary School, London.

The Mansford School, built in 1928, was the brain-child of the London County Council and an intriguing experiment. Enormous sums were lavished on equipment and facilities to teach technical subjects to working-class boys in the London area. The school had its own forges, machine shop, sheet-metal breaks, punch presses, lathes, drills, and drawing office—everything that today is taken for granted in technical-training institutions. In those days, Mansford stood out as a bright star in the technological firmament.

Although everyone agreed that Mansford was a splendid experiment, it proved too rich for the London County Council's budget. It was gradually phased out. But, by a stroke of good fortune, I benefited from the experiment and managed to get three invaluable years of personal attention from a brilliant mathematician named M. Maguire and an equally knowledgeable science lecturer, A. H. Raines. Without the instruction provided by these two men, I would never, I think, have had the theoretical and practical foundation that is essential to a career in telecommunication engineering.

My parents were not rich, although my father, a master tailor, earned a good living in the East End of London. He had emigrated to England from Pelots, a small town near Warsaw, in 1912, at a time when Poland was experiencing one of its periodic pogroms against Jews. England, at that time, was the chosen haven for East European Jews; the United States of America was second on the list, followed by Africa, South America, and, surprisingly last, Palestine.

My mother, Anne Harris, was a Cockney born and bred in Bow, although she, too, was of Polish-Jewish parentage. Her parents, who had settled in London from Poland in the 1880s, had changed their name to Harris from Schmidt at the outbreak of the First World War because of the anti-German mood then prevailing in Great Britain. My mother was a natural blonde, slim and of a quiet temperament. Given her family name and origin, I imagine there was a lot of German in her blood.

For my Bar Mitzvah, my father gave me an Avo Minor electrical-test instrument to measure voltage, current, and resistance. It was an essential tool for building and testing radio equipment. He also bought me a perforated Nipkow (or Nipkov) disc, a device that made possible the development of television and dominated early attempts to transmit moving pictures. No aspiring television engineer could work without one.

Cycling around the back streets of London searching for parts to build radio and television equipment, I made the acquaintance of Alan Blumlein. He ran a small electric-gramophone workshop in Petty France, Westminster, where he specialized in the production of high-fidelity music. He was a studious-looking man in his thirties with the air of an absent-minded professor. Narrow-faced, with large round glasses and straight, thick black hair, he was as thin as a beanpole.

Blumlein's operation was among the cluster of small workshops bought out by General Electric of the U.S.A. in 1931 when the giant corporation became EMI (Electric Music Industries) and built a factory at Hayes, Middlesex. In 1934, Alan registered a patent based on his negative-feedback circuitry, which led to what is now known as hi-fi. He also made 78-r.p.m. phonograph discs to be used with his new hi-fi equipment. After he joined EMI he became one of the firm's star radio scientists.

From 1935 on, I was a full-time service technician at the EMI Tottenham Court Road factory, installing and servicing radio and television equipment. This gave me invaluable experience in the field.

However, by this time, Alan Blumlein, the Michael Faraday of the electronics industry, was an old friend of mine, for I was a fourteen-year-old when I first met him. When I asked him to give a talk to the Chiswick radio club, of which I was a member, he agreed, in spite of a rather full schedule.

It was at our radio club that he predicted that stereo sound would one day become an international industry. By keeping in touch with Alan, I kept abreast of the latest developments in the industry.

During 1936, a frequent visitor to EMI's London factory was an RAF flight lieutenant who took a keen interest in our work. Because he wore civilian dress, it was some time before we knew he was an RAF type. No one was interested in the military in the mid-thirties, myself included, and as a result I knew him only by his first name, Bob.

Occasionally, Flight Lieutenant Bob lectured at the EMI factories at Hayes and Tottenham Court Road. At the end of one lecture he asked, "Anyone interested in joining the RAF?"

He must have thought we were stupid or something. We told him to get lost.

"Those who do," he said, "will learn to fly."

That was enough for me. I liked the idea of flying, but not enough to get into uniform for it. However, the upshot of my interest was that I began visiting the North Weald airfield and helped service the radio equipment in use at the time. The wooing was subtle and sure. Too late I realized that the air force needed radio technicians more urgently than it needed pilots, but by that time I was deeply committed to giving my help, part-time and unpaid.

In the summer of 1937, Flight Lieutenant Bob said, "Look here, old chap, what about giving us a little more help—part-time of course."

"What kind of help?" I asked.

"The ordinary stuff you're working on, cathode-ray tubes and high-voltage power units. That sort of thing."

"All right," I said. And so it was, in response to an official letter from the RAF, that I teamed up with two other volunteers from the London area to do unpaid work on the weekends for the government—at Bawdsey. One of my companions owned the battered Fiat Cub automobile in which we made the trip to Suffolk. We had come from London by way of Chelmsford, Colchester, and Ipswich, stopping along the way for a light lunch; light because our funds were limited, not because we weren't hungry.

I knew the driver slightly, for he was my main contact. His name was Bob Brown. The other member of our trio, Peter Friese-Green, had a strong North Country accent. There was little other than general conversation between us to make the long journey short, because no one had any real idea of what this was all about. We were unable to interrogate one another about the meaning of it all. Though none of us knew it at the time, each of us did have a valuable skill or trade.

We got a flat tire on the far side of Colchester and stopped to repair it, which cost us time. To delay us further, Bob,

5

who was supposed to know where he was going, lost his way. It was not an auspicious start to the weekend. The consolation was that I had Saturday morning away from my job in London. We were tied to a five-and-a-half-day work week, and this was a treat.

Eventually, though, the Fiat rattled noisily in the quiet of evening over the gravel, making an odd swishing sound. The sun, now sitting low in the west, cast a golden glow on the lush grass, and the shadows of the trees were fast growing to infinite length. There were no lights in the manor-house windows to indicate there was life inside the grey stone walls.

The house looked unfriendly and aloof, but no sooner had we parked in the forecourt and climbed out to stretch our legs than a heavy-set fellow crunched over the gravel to intercept us. His manner was brusque, though not unfriendly. We showed our letters of introduction and he grunted, "Oh! Orright then," and directed us to the front entrance.

As we approached, I noticed a motto in French worked into the mosaic floor before the weather-worn door. Translated, it read "Nothing ever changes", which was an ironic statement considering that in two short years the world would be turned upside down. Bob went inside to reconnoitre and soon returned. Someone had ordered him out. "I don't know. It seems a bit queer to me," he said, perplexed.

Moments later a tough-looking character with a flattened nose, wearing a tweed jacket that looked a size too small, came to the door and peered at us suspiciously. "What d'you want?" he demanded.

I offered an explanation and his face broke into a grin. East End English is easily identified by another East-Ender and is as good a passport as any in the right company. Still, he didn't budge.

Standing as impregnable as Dover Castle, he proceeded to give us a pep talk on security, speaking in the tones of the sergeant-major he had probably been in a former life. For a good three minutes, as we stood on the threshold, he hammered away at the need for security like Noah battering the last nails into the Ark, making us promise on oath not to discuss with anyone anything we saw at Bawdsey—wireless transmitters or the rest. We hadn't the slightest idea what he was talking about, but we asked no questions. I think we were all a bit scared of doing or saying the wrong thing.

"Well, never you mind," he said finally. "Just watch yourselves and mind what I say."

A picture flashed through my mind of the equipment I'd seen on the biplanes and other ancient machines at North Weald, and I remember thinking, he couldn't be talking about aircraft radios, surely! The aircraft radios with which I was familiar were unsophisticated sets, years behind the equipment I dealt with in my everyday work at EMI. All the same, the unsolved mystery of our journey, and the cloak-and-dagger atmosphere into which we had been plunged, filled me with trepidation. I resolved there and then to honour unto death the assurances I'd given.

The man who, in a manner of speaking, had read the Riot Act to us was named Joe Airey. Beneath his gruff exterior and foghorn voice was a capable engineer who could not be ruffled. He took us in hand like a father, showing us to our sleeping-quarters, finding us food because we had missed the evening meal, and helping us settle in.

The next morning, buffeted by a boisterous North Sea wind, we went to the Deben riverside, where a number of new huts were in the final stages of construction. The Deben flowed past the Quilter estate and into the North Sea. There was also a tall radio mast, and, near by, a cluster of half a dozen men were wrestling with an assortment of spars, trying to build another. I joined in; there was no one to tell us what else we should do.

By the end of the day we had erected a wobbly structure that, it seemed to me, might withstand a wind without collapsing, but not a gale. We three newcomers followed the others back to the house and went into the high-ceilinged baronial hall that served as the communal dining-room. We were all wind-blown, dishevelled, and tired from the day's exertions.

There were some inquisitive stares from those already seated at their evening meal. The more senior residents pulled our legs and wanted to know how we'd got out of infant school, but by the time the meal was over we knew we had been accepted and were part of the clan. Later in the evening, Joe Airey produced a list and ticked off our names. We knew then that we were truly in.

It is hard to convey a sense of the casualness of it all. At that time, as we had seen on arrival, security was virtually non-existent. During the late-summer weekends, local families visiting the shoreline wandered through the grounds, and the children climbed on to the masts and the piled timber as though they were in a playground.

A small group of scientists and engineers, working under the direction of Dr. Robert Watson-Watt, a brilliant scientist working for the General Post Office, lived and worked at the manor full-time. Watson-Watt, however, spent a lot of his time in London and elsewhere, but made frequent visits to Bawdsey to direct the research-and-development group, which included Arnie Wilkins, Ed Bowen, Bob Carter, and Don Priest.

There were six weekend volunteers, mostly radio men and non-technical personnel, including a photographer. Of the original group of twelve permanent staff and volunteers, nine were technical types who were specialists in the science of radio and television.

We were all young, none older than twenty-five, and, not yet being seventeen, I think I was the youngest. We were there to contribute our skills. What was not so clear is what we volunteers were supposed to do. All I'd seen so far was a manor-house, the new huts on the bluff near the river, some radio masts, and the other shaky structure I'd helped to build.

In the days and weekends of voluntary isolation that followed, there proved to be an immense amount of work to do. No matter what the task, whether it was making the place shipshape, doing carpentry, undertaking electronic construction, doing metalwork, or constructing radio masts, everyone lent a hand. In the beginning it was not very technical work, and there was certainly nothing secret about what we did. Yet this was my introduction to Air Ministry Experimental Station Bawdsey.

Of all the personalities involved in London, none were more important at the policy-making level than Sir Henry Tizard and Professor F. A. Lindemann (later Lord Cherwell), who, before the First World War, had studied together at Leipzig University. Both were former test pilots: Lindemann for the Royal Aircraft Establishment at Farnborough, Tizard for the RAF. Although they had been friends at university, their paths diverged and, in recrossing, a professional rivalry developed between them, especially on the subject of radar. Their relationship was touched on by R. V. Jones in his book *Most Secret War* (Hamish Hamilton Limited, 1978).

In November 1934, Harry E. Wimperis, Director of Scientific Research at the Air Ministry, had recommended the formation of a Committee for the Scientific Survey of Air Defence. Tizard was appointed chairman, Harry Wimperis was made his assistant, and A. P. Rowe was named the Committee's secretary. Other members were Professor A.V. Hill and Professor P. M. S. Blackett, all close associates and confidants of Tizard. The Committee's main task was to discuss what steps might be taken to improve Great Britain's air defences. Tizard had the idea that national defence against bomber aircraft was possible by electrical means.

The Committee asked Robert Watson-Watt to conduct a test of aircraft detection by radio waves with an Air Ministry observer in attendance. Watson-Watt was happy to oblige. Since the First World War, he had been working with his assistant, Arnie Wilkins, on complex radio-transmission problems for the Post Office, mainly international communication problems in the upper-atmosphere layer known as the ionosphere. (The ionosphere is a fluid layer of ion particles above the earth that reflects radio waves as a mirror reflects light waves. It is this phenomenon that makes long-range radio transmission possible.)

In a field at Weedon, just north of London, Watson-Watt had demonstrated to the satisfaction of Air Ministry officials that radio detection of aircraft was indeed practicable.

To further conduct his experiments, Watson-Watt was given a government grant of £10,000 to explore the new technology. He chose a small team of scientists and, from the Radio Research Establishment of the GPO (General Post Office), of which he was an employee, he brought his assistant, Wilkins. The tests, in January 1935, were conducted on Orford Island, a few miles south of Aldeburgh on the Suffolk coast.

In the January 1935 test, Watson-Watt used the BBC 13-metre Empire transmitter at Daventry. The experiment was to demonstrate that radio transmissions would bounce off a Heyford aircraft, flown from Farnborough, and register an indication of the aircraft's presence on a cathode-ray tube. The equipment was installed in a General Post Office Morris truck by Wilkins. Wilkins and the GPO driver spent the night in the truck near Weedon, Northamptonshire, ready for an early start the next day when Watson-Watt would arrive with Rowe of the Air Ministry to witness the demonstration. The test was successful. The aircraft was detected at a range of seven miles.

Winston Churchill had become involved in the whole question of air defence as early as 1934. He was not in the government at that time, but was asked by Prime Minister Stanley Baldwin to help in what was then becoming an urgent matter of national security. With the rise of Hitler and his National Socialist Party in Germany, there was justifiable apprehension in the rest of Europe about German militarism. It is a strange sidelight on British politics that the then Prime Minister should secretly request the aid of an unpopular member of Parliament. He did, however, and what is more he received Winston Churchill's whole-hearted support. Despite the oddity of his situation, Churchill wielded considerable power and influence—even when out of favour.

Professor Lindemann's position was no less unusual. He was Churchill's scientific adviser and was on public record in a letter to *The Times* in stating his firm belief that no one would find a way of detecting aeroplanes by electrical methods. When Churchill, always in the know, told him that a special air-defence committee had been formed, and electrical detection had been achieved, Lindemann, who had previously viewed radio detection of aircraft with scepticism, had second thoughts. He felt that he should be chairman of the committee, insisting that he, because of his experience, would make a far better chairman than Henry Tizard. He undoubtedly regarded the position in terms of its prestige, as a platform from which to exert his influence.

Churchill tried to have Tizard replaced by Lindemann but failed, mainly because the other members, loyal to their chief, threatened to resign if Lindemann were made chairman. For this reason he was able to sit on the committee only as an ordinary member. All the same, from that time on, according to some committee members, there was a strong feeling that the committee functioned less effectively than before Lindemann joined, basically because he sought to impose his own ideas on aerial defence. This wouldn't be surprising. He was a forceful advocate of his own opinions.

Lindemann, it seemed, favoured the creation of aerial minefields around industrial towns by the use of giant balloons with mines dangling from them on steel piano wire. He quickly became dissatisfied with the slow pace and characteristically phlegmatic approach of his fellow committee members. Obstructive in his attitude, as well as ambitious, he asked Churchill to suggest a more authoritative committee at a higher level, with himself installed as chairman. This would put him one up on Tizard. Nothing, however, came of this suggestion, and by then Watson-Watt's work on the coast of Suffolk was well under way.

The area on the Suffolk coast chosen for further experiments, Bawdsey Manor, east of Ipswich, was ideal for a number of reasons. For one thing, it was in a sparsely populated district with poor access to road traffic; for another, it was on the flight path between London's Croydon International Airport and the continent, so there was an unending stream of no-cost test flights passing the station, on which the newly developed electrical detection equipment could practise.

During the spring of 1935, the scientists and work-crew had lived mainly in the Crown and Castle, a small hotel in the village of Orford, Suffolk, in the shadow of Orford Castle. Many of the ideas and devices developed during the war were first discussed in the cosy lounge of the Crown and Castle. As the team expanded, a more permanent base of operations was needed, which is how nearby Bawdsey Manor came to be chosen. It was common knowledge at Bawdsey that, in the early days when they were developing ideas, the scientists left their drawings, notes, and sketches above the fireplace of the Crown and Castle. So it was not in a scientific establishment, or a boardroom or a drawing-office, but in an ordinary English pub that radar was first debated and ideas about it were exchanged. With better quarters at Bawdsey and improved equipment, Watson-Watt and his companions began assembling the pieces of the radar jigsaw and gradually transformed theory into practice.

Bawdsey lies a short distance along the coast south-west of Orford. The manor was the nearest reasonably sized building in the area and, since it was about a hundred feet above sea level, and since there was no ground in the area to cause obstructions, it was ideal for radio experimental work. Arnie Wilkins and Joe Airey, two of the team, were the first to discover the manor-house. After surveying the premises, Watson-Watt and A. P. Rowe, the Minister for State's assistant, approached the owner to sound him out on renting the premises to them. Sir Cuthbert agreed. He hadn't the faintest idea what the work was about, but he was patriotic enough not to ask difficult questions. Obligingly, he moved into one of the towers to make room for the new tenants.

By the time I arrived in 1937, fresh and eager to join the team as a volunteer, the experimental station was a well-established reality. My weekend visits to Bawdsey, with meals and accommodation provided, soon became a matter of routine. A few weeks passed before I met Watson-Watt. We were in the dining-room eating sausages and mash when he came and joined us. He was an Aberdonian with a round face and a jolly, outgoing personality.

Something about our plebeian meal must have struck a chord, because he made an impromptu speech in broad Scots, praising in an enthusiastic tone the infinite superiority of the haggis over the English sausage. This caused laughter, for he was an articulate speaker.

Watson-Watt had all the qualities that make a good leader: sincerity, humanity, wisdom, courage, and dedication. These were the virtues he needed to get the physicists, academics, and technicians to work together on the almost insuperable tasks he set them—and to get them to achieve. His methods were simple and direct. For example, he regularly organized intellectual free-for-alls so that, no matter who you were or what your position, you were allowed to have your say. Today we call such sessions "think tanks", whereas Watson-Watt, with his socialist background, called them "Sunday Soviets". These were held later at Swanage and Malvern, and were eventually attended by our most senior scientists and soldiers. "Soviets" were helpful in solving many of the problems we encountered.

One Sunday Soviet brain-child was the Cathode Ray Direction Finding (CRDF) system known as the Huff-Duff, which was to be instrumental in our winning the U-boat war. The same device would help us detect V-2 rockets near the war's end. There is no question that Watson-Watt invented the CRDF system. In 1943 he was, however, to be beaten to the punch by Canadian general Andrew (Andy) McNaughton, who registered the invention two months before Watson-Watt applied for a patent. With such men as A. F. Wilkins, R. Carter, W. A. Butement, E. G. Bowen, A. Budden, G. Touch, J. T. Herd, and D. H. Priest, Watson-Watt gave birth to practical ideas that helped British radar developments surge ahead of German radar technology and, consequently, were to enable the Allies to win the radar war.

I spent a great many weekends at Bawdsey, as well as holidays and, from time to time, even an entire work week. In the meantime there were evening classes to attend. In addition to electrical-transmission theory at the Regent Street Polytechnic, we had lectures from visiting specialists on a strange and eclectic variety of subjects. I also attended a four-lecture course on air navigation, for I was still hankering after that chance to learn to fly.

There were few, if any, radio navigation aids for aircraft pilots before the war. Navigation in the air was based on "dead reckoning", which was the art of guessing where you might be at any given time, provided you had accurately estimated wind and aircraft drift. The short course on navigation was to prove of immense help when I tried my first night-fighter control work.

My closest friends in the immediate pre-war years were fellow East-Enders George Smith and Tommy Baxter. The three of us spent most of our free time cycling and camping and—when not indulging in those worthy exercises—attending to the affairs of our radio club in Chiswick. Occasionally we listened to political speeches, despite the fact that we had no fixed political views except, perhaps, the faintly socialistic opinions of young men in search of a cause. I certainly wasn't conservative—not until, by chance, we heard Winston Churchill speak.

We were camping at Stones Farm near Waltham Abbey on the eastern outskirts of London one summer weekend. Hearing there was a political meeting of the Conservative

Party in the vicinity, we decided to go along and have a bit of fun heckling the speaker—whoever he might be. The speaker turned out to be Churchill, and who were we to heckle him? Exposed to his oratory, we were soon caught up in the enthusiasm of the crowd, clapping and cheering for all we were worth. I didn't equate my weekend work at Bawdsey with politics until that time, and what he had to say in clear and simple language planted the seed that would change my outlook completely.

His entire speech was taken up with Hitler and his gang, leaving us in no doubt as to the course of European history if nothing was done to resist them. Winston exhorted his audience to help him in his campaign against the evil that was taking over the continent. For days after the meeting I found myself wondering what ordinary people could do. I now began to directly connect the Bawdsey work with Hitler's designs on Europe.

I began listening more carefully to what my father had to say after the Churchill speech, for my father was clear in his own mind about where his loyalties lay. Having known persecution in Poland, having served with his brothers in the British Army during the First World War, and having been a fierce patriot in his land of adoption, my father was an outspoken advocate of British freedom. "This is the one place where people are still free," he would tell me. "If you have to choose between giving in and fighting, fight; just remember that. Fight with everything you've got."

All this he said without an inkling of what I was up to at the weekends. My parents never asked, and I never told them. They thought I was out cycling and camping. Joe Airey had told us to keep our mouths shut and I did. The letters RDF never passed my lips outside Bawdsey.

I realized that the Bawdsey experimental work was an important part of the national defence system and that I, a tiny cog in the machinery, was part of it. I worked hard to learn all there was to know about the new science and found the work exciting. Realizing that the atmosphere in which

15

we worked really was secret, even though in many ways mundane, had an effect on my sense of intrigue.

Shinning up the 250-foot-high radio towers like a powder-monkey, I learned a great deal about aerials and direction-finding equipment—and about climbing, which exercised the leg muscles. Indeed, my time was dominated by the theory and practice of radio, television, RDF, and quality-recorded music, which was being pioneered by EMI.

Considering the lead EMI took in the manufacture of high-fidelity equipment and the recording industry, many of the most senior engineers on staff took little interest in the strides being made in television, which, then in its infancy, was making great progress. For them, television was a passing fad that would fade away, and consequently they made no effort to study the subject. For this reason, by 1938 I found myself, when barely nineteen, teaching television theory to men twice my age. Radio- and television-wave theory was but a short hop from RDF technology, if they had only realized it.

The secrecy maintained about Watson-Watt's work on radar was absolute during the immediate pre-war period. It was nothing short of amazing that no word of our work was made public. There were no prying journalists to stir the waters, no leaked documents or rumours to pique the inter-est of investigative reporters, nothing in fact to attract the interest of German radio experts. In no other way would our work have withstood the scrutiny and probing of the "competition", which was to come later.

2

PRE-WAR BRITISH RADAR

Robert Watson-Watt had demonstrated in 1935 the feasibility of aircraft detection by means of radio. The test had shown that an aircraft would reflect radio waves and that this effect could be observed on a television-type screen. The problem now was to turn this exciting phenomenon into a practical device for the use of the RAF. In other words, the bearing, or direction, of the aircraft, and more accurate measurements of its speed and distance (or range), would have to be determined.

Prior to the invention of radar, so-called radio direction finding had been practised for years and every radio engineer thought he understood the meaning of the letters RDF. The canny Scot Robert Watson-Watt deliberately maintained the use of the old abbreviation to cover the innovative new research being done. The task at hand was to develop equipment that would be powerful enough to transmit a radio pulse up to 100 miles, and yet be sensitive enough to pick up the echo returned from the aeroplane.

Radio waves can be measured in two ways. First, there is the length of the wave, or wavelength, which can be visualized using an analogy, as, for example, the distance between the peaks of two succeeding waves. Second, there is the frequency. Frequency is the number of wave cycles that occur in one second. Frequency and wavelength are inversely related. For instance, a 300-metre wavelength corresponds to one million cycles per second (1 megacycle). This is a very low frequency for radio waves. Further down the radio spectrum, wavelengths between 60 and 25 metres are referred to as short waves or high frequencies (HF). Wavelengths in this band are used for overseas broadcasts. Transmissions that have even shorter wavelengths (down to one metre in length) emanate at very high frequencies (300 megacycles, or 300 mega-hertz), and are called, appropriately, VHF.

Beyond the VHF band are the ultra-high-frequency or UHF bands, which work at thousands of cycles per second. All television transmissions and most radars work in the VHF and UHF bands.

The basic principle of radar is fairly simple. The trans-mitter emits short, sharp pulses. A single pulse is transmitted from the radar transmitter and the receiver waits for the echo to return. Since we know that radio waves travel at the constant speed of 186,000 miles per second, we can com-pute the distance of a foreign object from the transmitter by measuring the time it takes for the echo to return.

The last important thing to understand about radar is the design of the aerial used to broadcast the signal. For this purpose, a suitable metal rod cut to the right length is used. The length of the rod, called a dipole, must be half the length of the wave to be transmitted. Therefore, the dipole for a 12-metre-long signal would have to be 6 metres in length. A single dipole will broadcast the signal in every direction. If more than one dipole is used to transmit, the signal begins taking the shape of an ellipse. With more dipoles, the signal becomes more concentrated, more narrow and directional. With a reflector, the signal can be trans-mitted in one direction, and, paradoxically, the greater the number of dipoles used, the narrower will be the transmitted signal.

All of these wavelength, frequency, and aerial or antennae phenomena are now known to radar and telecommunication engineers.

But radio broadcasting was still a relatively young technology in 1935. The BBC had, however, been broadcasting for some years on short wave to Australia and South Africa. To reach those distant places, it had to use a very short wavelength as well as very powerful transmitters, which were built by Metropolitan Vickers Ltd. Co-operation between the Post Office and the BBC had produced an enormous amount of experience in the design of massive aerials for high-power radio transmissions. In other words, some technology had already been developed, but Watson-Watt and his team of scientists were aware of the need to work at higher frequencies for which no suitable equipment yet existed. Furthermore, some ability in range determination had been achieved, but the direction (or bearing) of the aircraft was still a problem.

In the summer of 1935, leaving the smoky-black arches of Liverpool Street station in London to travel by train to Bawdsey, Watson-Watt had had an idea for attacking this very problem. His inspiration was to adapt an old-fashioned (1904) low-frequency radio direction-finding instrument for VHF service. It is ironic to describe instruments as "old-fashioned" in a technology that was barely off the ground—but old-fashioned it was. The instrument he adapted, a goniometer, was able to detect the direction of a radio wave. Watson-Watt's brainwave was to adapt it to detect VHF waves.

The system in use at that time (1935–36), called "range cutting", required the use of two RDF stations in separate locations to pinpoint the aircraft. The operators would draw two short arcs at the ranges read by the two stations, and where the arcs crossed is where the target was located. The system was cumbersome and time-consuming, but with the new arrangement using Watson-Watt's modified goniometer, it was possible by late 1936 for a single station to pinpoint an aircraft's position on a map, because the device could determine the point from which the signal "bounced" back.

Once a workable prototype unit consisting of a transmitter, an aerial, and a receiver had been developed, it was necessary to involve commercial manufacturers. The discipline of secrecy demanded by Tizard and the Defence Committee was absolute; this could have been a problem when commercial firms became involved in the project in 1937. To maintain secrecy, no single manufacturer was permitted to produce more than one item of the complex equipment: one company built transmitters, another the receivers. Other firms were contracted to make the auxiliary equipment needed to make a station operable.

By today's standards, the equipment was rugged, bulky, and large. The tubes used in the transmitters required high voltages for their operation, and the knobs and dials for adjusting the signals were crude. Altogether, the prototype installation looked more like the control panel of a steam-generating station than a radio station, but it worked.

The plan was to construct a chain of identical stations along the east coast of Britain facing Germany, which was to be known as the Chain Home (CH) system. The aerial masts of the stations were impressive and conspicuous, for they soared 360 feet into the sky and could be seen for miles. They carried fixed antennae—one for the transmitter, another for the echo-receiver.

The Chain Home system, as mentioned, operated on a long wavelength of 12 metres (a little over 40 feet). It was the waveband used by the British Broadcasting Corporation (BBC) for its Empire broadcasts; it was also the waveband used by the Germans for the broadcasts of Dr. Goebbels, Hitler's Minister of Propaganda.

Not everyone on the Bawdsey team worked on the Chain Home system, for other developments were taking place at the same time. Some of these are worth describing, beginning with Dr. W. A. S. Butement's Coastal Defence (CD) system.

Butement was no newcomer to the Bawdsey team. In

1931, working at the Signals Experimental Establishment at Woolwich, Butement and a colleague, P. E. Pollard, made a pulsed radio system using a 50-centimetre-long wavelength for detecting ships. The navy was not interested.

CD began life as a system to aid gun-laying under the name GL, standing, of course, for gun-laying. The first CD system was to be used to protect Britain's shore from bombardment as had happened during the First World War. The cumbrous equipment, secured over a huge wire mesh that was mounted on posts a couple of feet above the ground, had to be rotated as the gun it served was turned. But the height calibration of the radar equipment changed with every movement of the gun, making the height-finding apparatus impractical for aerial use. Its modification for use as a surface vessel detector was more successful, and it was the first precision radar.

By splitting the radar beam in two and balancing the left and right beams, Butement with his CD was able to get a high degree of accuracy. Before the war, Winston Churchill visited the site for a demonstration. A small vessel was picked up and the radar aerials focused on the ship. Churchill, reading the range, insisted that, at the range given and considering the prevailing conditions of visibility, he ought to be able to see the vessel with his naked eye. He squinted through a telescope mounted on the aerial-control column and searched for the ship that should have been directly in the line of sight. He puffed at his cigar and declared that there was no ship to be seen.

A few seconds later the demonstrators sighed with relief: the ship, viewed end on, slowly emerged from behind the telescopic hairline. The story sounds a bit far-fetched, but, although I was not present at the demonstration, I was to work on CD equipment later and found the account extremely accurate. One of the main drawbacks was the fast corrosion of switch contacts caused by the salt sea air. This corrosion fault produced a jitter on the cathode-ray-tube trace, which made it difficult to even see a target vessel. It was a long time before this problem was solved electronically rather than mechanically.

Another Bawdsey work group, under the guidance of Dr. Ed Bowen, concentrated on the development of an airborne radar system that could locate ships at sea. The Bowen team effort resulted in the Air/Surface Vessel (ASV) system. The difficulty with airborne radar was to make the equipment small and portable enough for use in aircraft. The available space in pre-war aircraft was small enough as it was. To fit in awkward and heavy pieces of radar was a juggling act, but it was done. When at last it was ready, Dr. Bowen, through his chief's contacts in the Admiralty, arranged to take part in a naval exercise.

The Admiralty agreed, but for some unfathomable reason cancelled the exercise because of inclement weather. Low visibility was just what Bowen wanted; any fool, he said later, could see ships from the air in clear weather. In any case, unaware of the cancellation, Bowen and Bob Hanbury-Brown took off from Martlesham Heath aerodrome near Bawdsey and headed south for the sea in the aircraft equipped with the prototype ASV detection apparatus.

The ships they were hunting were an aircraft-carrier, two battleships, and a cruiser on naval exercise off the south coast. By the time the two scientists set off on their quest the ships were enveloped in thick fog. Despite the fog, they located the aircraft-carrier HMS *Courageous* and the battleship HMS *Southampton* ten miles off Beachy Head. Jubilant with success, Bowen tapped the pilot on the shoulder to turn back and head for home.

The pilot had lost his bearings for a while in the fog, and had no radio transmitter-receiver, but he finally landed at a tiny south-coast airstrip. Using a public telephone call-box, Bowen contacted RAF Coastal Command in Whitehall and gave the location of the warships. The information was passed on and, no doubt, duly digested, because it was as startling a demonstration of the power of radar as that given at Weedon in January 1935. The days when naval vessels could hide in fog or man-made smoke-screens were over. The radar scientists had proved in the 1935 Weedon demonstration that aircraft could be detected. Now they had proved the same point with ships at sea.

Yet another pre-war development was the design and fabrication of the Identification Friend or Foe (IFF) device for installation in aircraft of the RAF, in both fighters and bombers. Designed by Don Priest and improved by Bob Carter, both of the Bawdsey station, the device was housed in an 18-inch-square grey box and was carried in the aircraft cockpit behind the pilot. It was a cumbersome unit for fighter aircraft with little cockpit space. The unit transmitted a signal to the radar operator via the CH screen to indicate that the aircraft under surveillance was "friendly". The IFF unit is known today as a "transponder" and is installed in all commercial and most private aircraft to identify the flight to the ground controllers.

Between 1935 and 1939, British radar scientists made enormous strides. In addition to the various types of systems developed, tested, and manufactured, a chain of radar stations was constructed along the British south and east coasts, from West Prawle on the south-west coast of England to Netherbutton in the Shetland Islands. The chain, with a station every few miles protecting Britain's European flank, was to be our insurance against surprise air attack—provided every station could be manned and made operational in time. What we didn't know was whether the Germans too had developed radar, and, if they had, how advanced it was. They certainly tried to discover what we knew about radar, as will be described in the next chapter.

Other writers can discuss pre-war espionage with greater knowledge and insight than I, although I did have personal experience of what I suspect was one German espionage effort, which occurred in the summer of 1938.

I had spent a pleasant weekend on the Brighton-Eastbourne coast, and was cycling back to London on the Sunday afternoon, when I caught sight of the tail-end of a long column of cyclists in the distance, travelling at a fast pace. I put on a spurt to catch up and came upon them, all neatly dressed in khaki-green shirts and, by English standards of the day, very short khaki shorts. I tried to overtake them, but, as there was a long, steep hill in the vicinity of Redhill near Reigate, I found the going hard and spent the next little while keeping my distance behind the last man.

The weather was warm, the sun hot. With sweat dripping from my face and on to the low handlebars, I kept my head down, but I glanced up occasionally to check that I wasn't losing ground. If I couldn't overtake the column, I thought, I could at least join it. I drew alongside the last man, who gave me a friendly smile before concentrating once more on breaking the back of the long climb. We crested the hill and plunged into a forest; then, once on the down grade, I had a chance to inspect my companion. His shirt was covered with badges like those worn by Boy Scouts. The words on them were obviously German, which I didn't understand.

He tried to speak, but his breathless English was poor. He signalled his inability to converse with an expansive, toothy smile. Still, we were travellers on the same road and there was companionship in that. Further on, the leader signalled a halt and I thought I might as well join them. We wheeled off the road into a tea-garden, where I accepted my travelling companion's invitation to tea on condition he'd permit me to pay. With the help of comrades whose English was better, the German boy and I discussed cycling, things of mutual interest to the young, and where his party was heading.

These sunburnt teenagers of my own age told me they had been on the road for almost two weeks, having toured through Belgium and the Low Countries. The trip through Kent and Sussex, they said, would bring them to the end of their holiday. They would cycle almost to London, then turn around and return to Germany. They enjoyed the English countryside and would have liked to stay longer. We laughed, we smiled, we joked a lot. They were a healthy, happy-go-lucky crowd who were enjoying themselves immensely. Most of them, it seemed, had already had some military training, because the badges they wore indicated proficiency in shooting, gliding, physical fitness, and other soldierly activities. Whether or not they were members of the Hitler Youth I never did discover for certain, although I now imagine they were.

After a break they remounted their machines and rode away, leaving me to laze among the flower-bordered lawns of the tea-garden with its rose-covered arches and shady trees. After a long interval, for I had plenty of time to reach home before sunset (I was still living with my parents), I wheeled my cycle on to the road and travelled at a steady pace. Considering my leisurely progress, I was surprised to come across the column once more at the top of a hill. Bicycles were strewn alongside the verge on what I thought was a rather dangerous blind bend, while the travellers were clustered together in groups.

Their leader, with two other men of the party, was attending to a tripod-mounted camera, making a great fuss about the photograph to be shot. One of the trio had his arms stretched out as though encompassing the view, and planning precise photographs. From the top of the hill they had a panoramic view of the patchwork countryside stretching towards London, glowing in the rays of the setting sun.

It did not strike me at the time that the pictures they were taking might be used for purposes other than filling the family album. Years later, when the Battle of Britain was at its height, I remembered this incident and realized that the Luftwaffe must have obtained a detailed knowledge of the English countryside from their touring fellow-countrymen. I wonder how many other cycling parties photographed England and the rest of Europe, and for how long.

All the time the development work was going on at Bawdsey, and from the time of my first visit to the station with Bob Brown and Peter Friese-Green, I had two jobs. During the week I worked at Tottenham Court Road at the EMI

25

radio and television service establishment. Most weekends I worked at Bawdsey, even though there was the occasional cycling trip such as the one on which I'd encountered the German cycling tour.

At Bawdsey I acquired a good all-round knowledge of what was going on, for I worked where I was most wanted, with Watson-Watt, Hanbury-Brown, Priest, and the rest. It was a fascinating experience listening to the debates of the scientific intellectuals. As might be expected, the question of what the Germans knew and what they might discover about our work was frequently discussed. Roughly, there were two schools of thought.

One group assumed that, because the Germans had good radio engineers and scientists, we could take it for granted that they monitored our transmissions. The same group argued that if the Germans had also developed radar, they would realize what our transmissions were.

The opposite group was utterly convinced that radar was our "breakthrough", and that the Germans were ignorant of our fabulous radar detection system. What supreme confidence some had. The possibility of detection did, nevertheless, spur us to discuss what precautions the Germans would take to counteract British radar before any air attack on the British Isles was begun. It was very easy to simulate interference on our receivers, which, for a variety of reasons, were susceptible to many forms of jamming.

There was a continuous-wave type of jamming, which displaced the screen image and caused it to jump violently up and down. Then there were what we termed "railings", these being vertical bars that could be made to drift across the screen like a heavy curtain. We were disturbed by the possibility that high-power transmitters in German bombers could transmit high-intensity "hash", or noise. As we proved in tests, this would be the most difficult type of interference for our radar operators to work through, but we developed filters to overcome this problem. As in the favourite game of bluff and double-bluff so widely described by espionage novelists, we needed to take counter-counter-measures, and this we did.

Ever since the Munich Agreement period of September–November 1938, tests on stations had been conducted and every foreseeable countermeasure and counter-countermeasure had been formulated and engineered into the Chain Home receiving and transmitting system. Following research and development, countering apparatus was mass-produced and installed in many stations.

The first protective system, equipment that enabled a change to be made to the operating wavelength once jamming began, was incorporated in the main CH stations. When the equipment was installed, the operator could, at the touch of a button, switch to a radio channel that was not affected by German jamming. However, not all CH stations had this ability.

In the pre-war planning, we expected any air attack to come directly from Germany. France and the Low Countries at that time seemed fairly secure, so the warning system was built on the east coast in the direct path of expected flights from the German Reich. Many east-coast CH stations had a choice of at least three spot frequencies; if the one in use was jammed, the operator could immediately try another.

Another pre-war Bawdsey development—the Mazda "AJ" tube—relied upon the fact that neither the "railings" nor the "noise" were synchronized to the CH transmitter. To put this excellent anti-jamming system into operation, a special cathode-ray tube was to be used, which did not respond to random, unsynchronized signals such as "noise" interference.

In spite of all the debate and argument at Bawdsey, for opinion swayed first in favour of one school of thought about British proprietorship of radar, and then of the other, the consensus leaned towards the sole-ownership view of the radar secret. It is strange to think how conceited we were in imagining that only the British had the genius to invent radar. The Germans suffered from the same mental block, as we shall see. There was no way of preventing the Germans from detecting and monitoring our enormously powerful signals, and we visualized German radio-intelligence operators studying our transmitted pulses on an oscilloscope. If that was the case, our secret would be out and the Germans would know that we had radar. Assuming this to be the case, they would prepare countermeasures, and all our efforts would go to waste.

By the time Prime Minister Neville Chamberlain returned from Munich on November 1, 1938, waving a piece of paper and announcing "Peace in our time", the die was cast. Although few people in Great Britain realized it, in the RAF at least there were no illusions, and the radar stations that had been completed were being manned round the clock for the first time from September 1938 on. This was the period that coincided with the September 30 Munich Agreement and Germany's takeover of the Sudetenland the following January, which is why I refer to it as the Munich period.

The CH system was prepared. Those who leaned to the German-scientists-would-know view assumed that once the bombing began in earnest (and what bombs are not in earnest?), German jamming would be switched on to high power. Then, despite our ambitious preparations, we realized that our stations could be blinded if the Germans used the right tactics and equipment. The devices needed to jam the equipment, we knew, were simple to make if our operating characteristics were known. If those jamming devices were carried in just a few of the attacking bombers, the effect would be disastrous. A further danger was the fact that CH radar was practically blind below 3,000 feet. Butement's CD was modified, however, to be able to detect low-flying craft. After the sinking of the *Royal Oak*, CD was applied to the U-boat problem (called CDU; the modified CH stations were renamed CHL, for Chain Home Low Flying). Naturally, not until after the war, when I was able to speak to German radar experts, did I get to know the full story. British fears of discovery of their own radar secrets were real, although, as I have tried to convey, speculation about what we thought the Germans might know was certainly mixed. Nevertheless, comparison of the independent developments of the two protagonists produces a fascinating juxtaposition.

We have seen where the British were on the eve of the outbreak of war; where were the Germans?

3

PRE-WAR GERMAN RADAR

From the very beginning of Hitler's rise to power in the early 1930s the Luftwaffe was regarded by German military commanders as a kind of long-range gun that lacked an accurate gunsight. In 1933, Dr. Hans Plendl, a specialist in radio communication, was asked to design a system that would allow German bombers to attack distant targets with precision.

Lufthansa, the German national airline, headed by Erhard Milch, commissioned the Lorenz Company to develop a system that would allow air liners to make a "blind approach" to their destinations in poor weather conditions. The result was the Lorenz beam system, commissioned and in service by 1937. Based on the blind-approach beam idea, Plendl and the Telefunken Company designed a bombing system that enabled bomb aimers to release their bombs to hit accurately within 500 yards of the target, at a range of 200 miles by day or night.

Both this system, called the X-Gerat, and its predecessor, called Knickebein, made use of accurately projected radio beams. The aerials for projecting the beams were mounted on turntables located on Germany's borders in such a way that two separately transmitted beams could be made to intersect at any point in the British Isles. Tested by Luft hansa, the system was fully operational by the time war broke out.

To carry out the directions of the radio navigational-aid system, Lufthansa had a superb fleet of aircraft that could be quickly converted into military transports and bombers. Under the direction of General Ernst Udet, a First World War fighter pilot, Germany produced the Junkers, the Heinkel, and the Dornier, to name just a few of their aircraft, and at the beginning of the war many of the prototype German bombers could easily outfly and outrun British fighter aircraft.

I must point out that the so-called bombing beams were navigational aids to aircraft. They were not radar devices. The X-Gerat-system transmitters broadcast a dot-dash radio signal which became a steady tone when the aircraft was on the correct track. The aircraft navigator heard the signals on receiver headphones that received them in the same way that an ordinary AM radio picks up radio broadcasts. Radar, in contrast, picks up transmitted signals that are reflected back by a flying, moving, or stationary "target" object.

The reason for introducing the Knickebein bombing beam is that it is understood by many to be part of the radar story. A feature that the X-Gerat system did have in common with pre-war German radar was the accurate directional projection of the signal transmitted. This achievement was far in advance of anything the British had developed, although a rudimentary directional signal was used by Butement for the CDU system.

Because the Wehrmacht commanders were planning on taking the offensive, defensive devices such as radar had low priority in the German order of things. For this reason, two amazing defensive inventions that they had developed in 1935 were not mass-produced until 1939, shortly before Germany overran France and needed a defensive warning system against bombers of the RAF. One was an early-warning radar system manufactured by the Gema Company and code-named Freya. It was much the same as Butement's CD/CHL. With the rotating aerial mounted on a tracked vehicle, the unit was completely mobile and, once set up, could "see" aircraft within a radius of 75 miles, provided they were flying at a reasonable height. The enormous aerial, very much like a rotating billboard, was necessary, as this was the only way in which radio experts could project the fairly narrow beam required on the operating wavelength of nearly 3 metres. The transmitter acted as a radio lighthouse, the aerial being rotated on its platform. The larger the aerial, the narrower the beam, which gave precision in location-plotting. There was obviously a physical limit to the size of any aerial of this type, and one can imagine the tricks the North Sea gales played with it.

Freya was the early-warning system that picked up aircraft at long range, and therefore needed a high-power transmitter. It operated on a fairly long wavelength, as this was where the Germans could generate high power. The longest wavelength used in pre-war German radar was 3 metres, and this limit dictated the large size of the aerial needed to produce a narrow beam. Conversely, the shortest wavelength on which a system would operate, they believed, was ½ metre; this was because the tubes available simply would not amplify signals at a shorter wavelength.

Aware of this limitation, German scientists believed there were no shorter wavelengths upon which radar could operate. They had established the limits and were prepared to accept them as scientific dogma, an assumption that was to mislead them when they attempted to probe British radar.

The second German defensive system, designed and developed by Telefunken, was built to use the ultra-short wavelength (½ metre) and was given the code name Würzburg. Both the Freya and the Würzburg systems were triumphs of inventive genius and engineering. The Würzburg had a very narrow beam and was able, with precision, to direct guns and searchlights, for which it was mainly used, but its effective range was short, a mere 12 miles.

To the German high command in 1939 its most dangerous opponents were of course France and Britain. France, the German General Staff reasoned, could be dealt with in a land battle. The British Isles, isolated by the English Channel and the North Sea, would, if necessary, be beaten into submission by bombardment from the air, and then invaded. To deal with the RAF in the event that the British were to enter the coming war, they boasted of an enormous fighter force and 5000 bombers. The Germans had no doubt about the accuracy of their bombing beams. Their only worry, therefore, was Britain's ability to detect the Luftwaffe bombers.

The Luftwaffe Chief of Radio Telecommunications was General Wolfgang Martini. If any attack on Britain failed because of their good radar defence, criticism would be directed at him as head of the service. Because the scanning range of the German radar-defence warning system was so short, Martini and his colleagues were convinced that British radar transmissions, if indeed they existed, could not be detected from as far away as Germany. The problem was how to get a well-equipped and stable radio laboratory near enough to the English coast to conduct tests. The obvious answer was an airship.

High in the sky, with its engines switched off, an airship could remain hidden in cloud off the North Sea or Channel coasts and conduct surveillance tests in perfect security. Lufthansa pilots on scheduled commercial flights from Europe to Britain were told to keep their eyes open for any unusual aerial systems. Martini realized that no equipment could be manufactured to counter a possible British defensive system unless the radio frequency on which it operated could be discovered.

Germany's airships, the Zeppelins, had been taken out of service and placed in storage following a number of disastrous accidents, notably the fiery loss of the *Hindenburg* in 1937 after a flight to New York City. The Zeppelins, in Martini's opinion, however, were Goliath-like machines unsuitable for surveillance services because of their size. He asked for the production of two simple, blimp-type airships. The request met with opposition from the military-industrial controllers of Germany's war production because of more demanding war needs. German armaments production had a voracious appetite; finding materials, always in short supply, was a continuing headache for the war planners. Martini explained the urgency of the situation to Goering and Milch: it would be Goering's own pilots who would suffer the consequences of any British radar system that was not fully understood and for which no countermeasures had been prepared. Goering yielded, but to avoid squandering needed material in constructing a suitable craft he had two Zeppelins brought out of retirement and overhauled. This was an expensive procedure, but less costly than designing and building a new, smaller airship.

Following overhaul, the Zeppelins were equipped with suitable battery-powered radio receivers, the batteries being charged by the outboard-propeller-driven generators. The first test to prove the equipment was conducted in March 1939 with Martini on board. Equinoctial winds buffeted the 770-foot-long airship like a cork on a turbulent ocean, giving passengers and crew a rough crossing over the North Sea. Following a short stay in the clouds, during which some listening was done, the flight returned to its base at Mannheim.

For the first extended investigative flight, the expedition left Mannheim one early morning in July 1939, battling a strong head wind until the craft stood off the English coast, not far from Bawdsey. There the engines were switched off, and for a short time the great machine glided forward through heavy cloud before drifting south. With the radio receivers switched on, the engineers listened for evidence of British radar transmissions.

Despite the excellence of the radio equipment, all that could be heard was violent 50Hz (hertz) interference, which came as a deafening low-pitched hum. The engineers knew that the British power grid was a 50-cycle (50Hz) system that criss-crossed the countryside. Was there perhaps some shoddy electrical-engineering work on the power cables that resulted in this radiation of 50Hz radio noise? They decided that this was most probably the case.

Finding no fault in their own equipment, they moved north up the east coast towards Scotland, repeating the tests at various points with the same result: that strong and inexplicable interference. On their return to Frankfurt, the receivers were again tested, and modified for greater sensitivity at the frequencies under scrutiny. The wavelengths with which they were concerned were those, of course, in the ½- to 3-metre band, because, as far as they could determine with the technology known to them, this was the range of wavebands on which any radar would have to work.

Later in July 1939, another test was conducted. The results were as disappointing as they were on the first mission. When the second mission returned, Martini was concerned. The fact is, all Zeppelin flights began their search for evidence of British radar off the coast of Suffolk, in the vicinity of Bawdsey, so it can be assumed that General Martini had some knowledge of the existence of the Bawdsey establishment. This is speculative, of course, but it is a reasonable deduction all the same. Whatever the facts of his intelligence knowledge might have been, Martini ordered a third test, which was conducted on August 3, one month before the declaration of war on Germany by the British government. Again the investigators drew a blank. There was only that interference at 50Hz, and no evidence of rotating beams like those used by all German radar stations.

With the flight on August 3 and its disappointing result, there was still no definite evidence of a British radar system, and only confirmation of a given wavelength and operating frequency would have made it possible for the Luftwaffe to produce suitable jamming equipment.

To understand the next act in the drama it is necessary to again touch on some basic radio-wave characteristics, and for this we need to go back to the days when the Vikings were on plundering expeditions around the foggy, treacherous shores of northern Scotland. In an effort to avoid the rocky shores, which were often shrouded in mist, the Vikings practised a form of Stone Age radar.

A sailor, stationed at the mast-head, would give a short, sharp shout from time to time. He would listen for the echo, the delayed return of which was an indication of the ship's distance from the shore. Obviously, the longer it took for the echo to return, the more distant were the cliffs and rocks. By calculating the interval between the shout and the returning echo, he could estimate the ship's approximate distance from shore.

In radio terms, a short, sharp shout is called a "pulse"; radio pulses are emitted in only a few millionths of a second, and their travelling speed is known. By broadcasting, for example, one pulse every second and then electronically measuring the time taken for the echo to return, an operator could calculate the distance in miles of the object from which the echo had bounced. The object of the radar echo, of course, was a target aircraft.

Furthermore, continuing with our Viking analogy, there would have been no point in the helmeted warrior's repeating his bellow until he heard the echo from his last shout. The same applies to radar; one had to "wait" for the radio-pulse echo to return before the next pulse was transmitted. If the aircraft was a hundred miles away, this would take about $1/1000$ of a second. This is why the Freya radiated 1000 pulses a second: a detection range of 100 miles gave enough time to prepare for the relatively slow-flying aircraft. The Würzburg, with its short 12-mile-range requirement, had a fixed pulse-repetition frequency of 3500 per second. The British system, as already mentioned, used a much longer wavelength, and a much lower pulse-recurrence frequency (50Hz). It was this that deluded the German snoopers, who were looking for a pulse-recurrence frequency of 1000 needed for long-range detection.

The use of "blimps" for the tests would have been more successful than the stately, if ponderous, Zeppelins, for the Zeppelins acted as aerial systems that absorbed, like giant sponges, vast quantities of energy from the CH transmission that saturated the on-board receivers. With many watts of energy floating around the airframe, saturating it and charging it like a storm cloud, there was little hope of the investigators ever making sense of the signals.

In authoritative military and official war histories in which the subject is discussed, reference is made to the radio "interference" that marred the German search for British radar. This is the first time that the real source of the interference has been explained. General Martini's scientists were misled by what can only be described as tunnel vision. In fact, radar was, and is, susceptible to many forms of interference, a practice we were to develop to perfection ourselves in time for D-Day.

Because of the oversight of German radar specialists, Britain was spared a concentrated effort to disrupt its radar facilities during the crucial days of the Battle of Britain in the summer of 1940. The network was attacked, but not successfully. It is of major significance to the course of world history that RAF Fighter Command, left time and again with no fighter reserves, had to rely implicitly on the country's radar network to make the best use of its dwindling resources. It can be confidently assumed that, with the British radar network jammed, the RAF would have lost control of the battle and been defeated. The British Isles would immediately have been invaded. An initial plan for the appointment of District Gauleiters together with a primary plan for the occupation of Britain had already been issued to the German Security Service (Waffen SS).

Summarizing the differences between German and British radar developments, the Germans had advanced VHF and UHF equipment; they had perfected the technique of directing signals, both with radio navigational aids and with radar; but they dogmatically clung to the notion that radar wavelengths were limited to between ½ and 3 metres.

The British had taken a different approach. They came late to the knowledge of how to direct their signals; they were far behind in the technology to make high-powered VHF transmitters at even the Freya wavelength of 3 metres. Instead, they chose the 12-metre wavelength for the Chain Home system, their basic warning network. Using the 12-metre band, they could use British Broadcasting Corporation-type transmitter equipment, which was readily available. Actual direction-finding was carried out on the receiver aerial and, of course, the goniometer.

Lastly—and what confounded the Zeppelin snoopers—was the continuous 50Hz interference that was detected everywhere. This occurred because the CH radar did not use a narrow, lighthouse-beam type of transmission, but a 120-degree floodlight of pulses.

There were pluses and minuses on both sides. The remaining major difference between the two systems was the frequency at which transmissions were broadcast. Because of the inability of the British to build a high-powered frequency generator to drive the signal pulses, they relied on the 50Hz power supply of the British power grid. The Germans, with their greater technical ability, opted for the higher pulse-recurrence frequencies to get the maximum number of echoes from the target aircraft, which produced a clearer picture on the monitoring screen.

The British couldn't pick up German radar signals; the Germans mistook the British signals. As a result, both sides preserved the secrecy of their systems until the eve of the Second World War.

During the five years immediately preceding the outbreak of war, astonishing advances were made in scientific developments, but mainly in Germany. This was not just in radio communication systems, but also in such things as aircraft and weapon design, tanks, ships, offensive and defensive systems, nerve gas, and optical sights. Because so much of what happened is pertinent to this narrative, it is worth reviewing the situation. It is also necessary to summarize the arsenal of secrets if one is to understand the energy with which Germany prepared for total war. In retrospect, there is no doubt that German scientists and engineers took a commanding lead, for they were directed by a centralized organization, with clearly defined aims and objectives.

At Rechlin, on July 7, 1939, a few weeks before Great Britain declared war on Germany, a number of secret devices developed for use by the German armed forces were demonstrated to Hitler and his entourage. A prototype jet fighter that was put through its paces had short endurance but achieved a phenomenal speed and was of revolutionary design. A Heinkel III, equipped with rocket-assisted propulsion for take-off to increase its bomb-carrying capacity, was flown past. Dr. Plendl demonstrated the Knickebein system for blind bombing; models of the Würzburg and Freya systems were shown; a Messerschmitt 110 on display, fitted with 30mm cannon, was far superior to anything the British had in the air; and, as a finale to the show, a flight of two German fighters, the ME 109 and the Heinkel 100 (the 109 had recently taken the world land-speed record from the Heinkel 100) "shot up" the reviewing stand.

Hitler was not planning to fight the British until 1942, if he had to fight them at all, so the July demonstration must have given him a feeling of immense superiority. It might also have increased his confidence, if more confidence he needed, to make his next territorial assault, which was to be Poland. The remaining obstacle in the planned invasion of Poland was Russia, and even this was overcome when Foreign Minister Ribbentrop finally sealed Poland's fate by returning from Moscow with the Russian-German non-aggression pact. The invasion of Poland was mounted on September 1, 1939. Two days later Great Britain and Germany were at war.

4

THE OUTBREAK OF WAR

September 1, 1939, the day Germany invaded Poland, was a bright, cool autumn day. I arrived at the factory for the morning's work in a jaunty mood, for I was planning to visit Bawdsey that afternoon, and the Suffolk coast would be a good place to spend the weekend.

The rear of our workshop overlooked Gower Street and the Architectural Faculty of London University. The students with whom we'd always maintained a none-too-friendly "Saturday Feud" were absent from their normal shouting-posts. Instead, from across the narrow street came the guttural voice of Hitler against the background noise of Teutonic music—"Thump! Thump! Thump!" Stretching to peer through the high windows of the Architectural wing, we could see our adversaries clustered about a blaring radio.

"This is it," said Freddy in triumph.

Chief Engineer Freddy Pankhurst, my direct supervisor, was a large, tubby man, who smoked a pipe and used the stem to emphasize his strong and sometimes bigoted opinions, for he was as outspoken as his more famous namesake, Sylvia Pankhurst.

"What is it?" I asked.

"Germany marching into Poland. It's been going on all morning. Don't you ever listen to the news? We'll have to get stuck into the blighters again, mark my words." Freddy was an older man. He had seen service in the First World

War. I didn't think he would have sounded as exuberant if there had been any likelihood he'd be involved in this one.

"Well, Jack, what are you going to do? Join the Salvation Army?" he asked, billowing clouds of pungent smoke from his pipe.

"We'll see. We'll see."

"Make up your mind . . ." he said, ready to launch into one of his bellicose lectures, but I wasn't listening.

My spirits had fallen, because I was convinced that our chances of winning a war against Germany were remote. A lot of people must have thought that way when they heard the news, but I think I had more reason to feel depressed. Spending my weekends at Bawdsey, I picked up a lot of information as time went on from the chit-chat that flew back and forth, and I had more to base an opinion on than many people. Though Freddy and I had often discussed air defence, I never talked about the Bawdsey work with him; the letters RDF were safe with me. Once or twice he'd prophesied that "our people" would find a way to "echo sound" aircraft by means of radio. I couldn't tell him how right he was, and not for many years did he learn of the work I had been doing.

Of the twenty-five Chain Home stations completed and covering the south and east coasts of Great Britain by September 1, 1939, fifteen were manned and in operation. The others, fully equipped and operational, were not manned, while the groundwork was complete for many more stations in the chain. The difficulty of operating the completed stations was of course the lack of trained personnel. Almost a year previously, an SOS had gone out to the major educational centres for personnel to staff the stations. Given the unusual and secret nature of the work, and the inability of the authorities to say just what sort of qualifications technical candidates needed, this was not an easy bill to fill. Many months were to go by before all the necessary stations were built, equipped, and staffed.

Fighter Command Headquarters was located at Bentley

Priory, Stanmore, an elegant suburb of north-west London. Fighter Command HQ was linked by an extensive network to every Fighter Group headquarters, to an army of ground observers, and to every completed and operational CH radar station. Each CH had its own WAAF (Women's Auxiliary Air Force) plotter assigned to it, who with a croupier's stick pushed the plaques around the table upon which the linear military grid map had been painted. At Group HQ, meanwhile, plaques representing Fighters, Hostiles, and unidentified aircraft were similarly punted around the table under the watchful eyes of the Duty Controllers representing each command of the air force. Coastal, Bomber, Ferry, Fighter, and Fleet Air Arm Controllers sat high on the balcony at Stanmore ready to claim ownership of any new tracks that appeared on the table.

Raids or unidentified tracks together with Hostiles were handed over to the appropriate Group for action. Group, in turn, instructed the appropriate Sector station by telephone to carry out the interception. The Fighter Sector stations had squadrons assigned to them, many out at satellite aerodromes, and of course they directly controlled the fighters. In this way the co-ordinated Fighter Command prepared to defend Britain against air attack.

There were times when more than one station plotted the same aircraft. This was eliminated by setting up a filter table at Stanmore, where Filter Officers, specially trained for the purpose, decided if several different tracks were of the same aircraft. This arrangement, whereby all data was processed through a central control, meant that discrepancies could be sorted out quickly, and fighter response be made most efficiently. All tracks were finally plotted on a large General Situation table map, and the information that concerned a particular Group was passed by telephone to the Duty Controller.

There was no direct fighter control from the radar tube at that time. A radar operator could not contact an aircraft. Therefore, sector control was at times ineffective because

42

delays often occurred in passing information. The whole system centred on the CH network and the Observer Corps, a civilian group mainly concerned with overland aircraft reporting. Early warning depended on the very-long-range pick-up of aircraft by the radar stations so that the whole system could swing into operation—and, as noted, many of these stations were still only in the planning stage.

Since I knew the state of our radar warning system, my apprehension was understandable. It seemed that our only protection would be the gas mask issued to everyone, man, woman, and child. The possibility of gas attack was cause for alarm, and rumour had it that the masks issued were ineffective against a new German gas called Arsene. We were told that if, after a raid, we saw wisps of cobwebby, silky thread, it would be the end. In fact, it was discovered after the war that the Germans had developed and stocked large quantities of nerve gas, which Hitler decided not to use because he was convinced that we, too, had stocks of nerve gas. He was wrong.

I made discreet inquiries as soon as war was declared and was told that a headquarters for all RDF work was being set up somewhere near London. It wasn't long before I was to learn more, for soon after I was sent to the RAF station at Uxbridge in the western suburbs of London.

While the authorities were deciding what to do with me, I had my first taste of life as an aircraftsman (A/C), which was not a pleasant experience. There were straggling line-ups for meals in ice-cold weather. We waited huddled in our overcoats in drizzling rain while the line slowly inched forward; there were miserable hours spent sitting on barrack-room beds being bored to distraction because there was nothing to read; there were cross-country runs in shorts; and there were interminably drawn-out parades. It was a depressing experience after the hustle and bustle of civilian life.

The one bright spot of it all was meeting up again with the comical Peter Friese-Green, my Bawdsey colleague. We commiserated with one another and, as much as we could, kept our own counsel. Then, one wintry morning in late February 1940, we were called off parade and driven to Oxenden House in Leighton Buzzard, a few miles north of London. That car ride to Oxenden House was luxury indeed after the bleak life of the Uxbridge camp. Leighton Buzzard is a short distance from Bletchley Park, which was the assembly place for us and the many other specialists gathered from far and wide.

The Park was the site of the Codes and Cyphers Headquarters where German Enigma communications were decoded. British Intelligence had obtained an Enigma machine from Poland before the country was invaded. The encoder machine, used by the Germans for their most secret communications, was thought by them to be unbreakable.

Oxenden House was a Victorian mansion, sparsely furnished, its walls decorated with ornately patterned wallpaper. The spacious grounds of the property were well kept, the grass was cut short, and the avenues of poplar, oak, and yew were a pleasure to walk in. There was little else to do but walk and talk and relax. After Uxbridge, we felt like good-conduct prisoners on parole.

All in all, we were comfortable in our new surroundings with good food and warm beds while we awaited the arrival, so rumour had it, of the newly appointed station commander. The people gathered in were scientists, academics, and specialist technical types like ourselves, as was evident from the discussions that took place. It was like the gathering of undergraduates waiting for the next semester to begin.

No one discussed radar: there was total silence on the subject; nor did we discuss the subject in the presence of others. If anyone else had radar experience it would have been hard to tell, for I recognized no one beyond my two Bawdsey companions.

Finally, one morning following days of idleness, word went round that the new station commander had arrived.

The commander, Air Commodore Gregory, lost no time in organizing us. Shortly before noon on the day he arrived we were herded into an assembly hall for an address.

He was a distinguished-looking man in his late fifties, an administrator with an engineering background, from what I could gather. He was a genial fellow with a quiet, scholarly manner and he had the right personality to organize an army of boffins. (Boffin was the wartime name given to any scientist, engineer, or technician who worked in the "back room".) His measured, gentlemanly attitude was in sharp contrast to the trumpeting of the Uxbridge NCOs with their foghorn voices that bid us "Get fell in!" From him, we learned that we were to be part of the newly formed RAF 60 Group, and that Oxenden would be the Group's headquarters. He explained that the Group was responsible for all radar stations: their construction, their staffing, their training, and their administration.

In his address, he spoke quietly, but with force and conviction. Describing the work to be done, he stressed the need for intelligence and logic in all we did, for communications was a new scientific field that needed to be explored. He told us that we had been chosen for our knowledge, skill, and experience, and would have to use all our intellect to overcome the problems with which we would be faced.

It was an inspiring speech that gave us confidence in ourselves. I know I felt stimulated and I'm sure others did, too. We were made to feel like the members of a gentlemen's club that had a strict code of behaviour, yet that demanded our complete allegiance. And this is exactly how I felt about 60 Group throughout my service. A return to our respective units was necessary, he explained, to learn some basic facts about the RAF. After his introductory talk, which put us "in the picture", we felt that we could tolerate the indignities of the disciplinary instructors of Uxbridge if only because we knew that we would soon escape the regimen, and that is what happened.

We remained at Oxenden for a few days following

45

Gregory's address, being organized and getting to know one another. We two from Uxbridge were then driven back to our unit in style, like members of an elite company of adventurers, reconciled to the fact that we must assume the guise of lowly erks (A/Cs) for a while longer. Within the week we were sent to the seaside resort of Morecambe, to be taught the ins and outs of RAF lore, and two weeks later, without further ado, we once again entered the secret embrace of the radar world. While I might bemoan my fate at this early period of the war, things were happening on the radar front which I learned about only subsequently. A most important development was the Oslo Report.

No discussion of the radar war would be complete without mention of the Oslo Report, so called because this comprehensive document, emanating from German scientific sources opposed to the Nazi regime, came into the hands of British Intelligence by way of Oslo. The Report has been dealt with at length by R. V. Jones in his *Most Secret War*. Part of the document dealt with the German bombing beams and radar.

I do not wish to imply that I was at any time privy to the Oslo Report, although the radar information it contained filtered down to us in time. Two types of radar were mentioned. One was an anti-aircraft-gun control; the other dealt with the detection of hostile aircraft. Details were sketchy, but it seemed to prove that the Germans had radar. The report even gave the range performance of the two systems which we later recognized as Würzburg and Freya. It was obvious from the intelligence that we were in a race to perfect our radar devices and systems.

We thought we were well served by what we had—the Chain Home stations, the IFF (Identification Friend or Foe), the ASV (Air/Surface Vessel), the CDU (Centre Detection U-boat), and the rest—but what *exactly* did the Germans have? A substantial part of the radar community still thought our bag of tricks was comprehensive. This was far from the case, despite our best efforts to perfect the equipment by trial and

46

error, and by native ingenuity. In retrospect, I can say that the equipment and systems then developed were to modern radar what Henry Ford's early Model T Ford was to the modern automobile. And, like the Model T, our 1939 radar suffered from some major weaknesses.

Almost immediately after the outbreak of hostilities, the Germans set in motion a plan to close the Port of London to shipping. A glance at a map of the Thames Estuary will show that all ships moving in and out of the port must pass through a narrow channel some five miles wide, bounded by the town of Shoeburyness on the Thames's north shore and by Sheerness on the south shore. The estuary was and is the main artery of London, through which most of its life-blood flows. Seal this gap and the Port of London would be cut off from the sea.

Germany had invented the magnetic mine near the end of the First World War. By 1939 it had developed a mine that could be dropped by parachute. For a few weeks, starting in mid-October 1939, a steady stream of Luftwaffe flying boats from air bases on Sylt and Borkum crossed the North Sea to mine the Thames Estuary between Shoeburyness and Sheerness. The effect on the estuary was immediate, and the results were catastrophic. The narrow waterway was soon clogged with sunken ships. RAF Blenheim bombers were sent to bomb Sylt and Borkum, but were not successful in reducing the number of mines dropped into the estuary. Something else had to be done.

The obvious solution would be to plot incoming aircraft to locate the probable positions of the parachuted mines. CH Canewden was on the estuary, but CH radar was unable to reliably detect aircraft flying under 3000 feet and certainly not at sea level. A system for plotting aircraft flying under the "net" was needed. The problem was put to Dr. John Cockcroft, who reasoned that the answer lay in a variation of Butement's CD/CHL equipment.

For accurate plotting, the aerial had to be small enough so that it could be rotated by hand. This meant using the very short (for us) wavelength of 1½ metres. The CH system, remember, worked on the 12-metre wavelength. Cockcroft had a model in Dr. Butement's CDU units. In fact, he had just returned from the north of Scotland, where he had been installing a number of CDU stations for detecting U-boats. The sinking of the battleship *Royal Oak* in our naval base at Scapa Flow in the Orkney Islands by a U-boat, *U47*, commanded by Lieutenant Günther Prien, on October 14, 1939, was a severe blow, and it was hoped that CDU radar might prevent a similar disaster in future.

By using eight half-wave dipoles (a dipole, as earlier explained, is an aerial rod half the length of the radio wave being transmitted) side by side, and stacking them in four sets of two, Cockcroft narrowed the beam width to 12 degrees, and succeeded in making a pencil-thin beam that could be directed to any point of the compass. By concentrating the beam, he achieved a major breakthrough for British radar and, by making it directional, the beam could be operated down to sea level. The design was similar to the German Freya and Würzburg units.

A large net of chicken-wire behind the dipoles served to reflect the beam forward. The aerial assembly was mounted on a wooden frame, 60 feet long by 20 feet wide, supported on a 6-foot turntable. To rotate the assembly, a sprocket was fitted to the centre shaft and connected by a long bicycle chain to a bicycle crank. The crank was mounted on an inverted frame set in concrete. To operate the device, the "binder", as the operator was called, sat with one leg on either side of the bicycle frame and turned the pedals by hand. The "bird-cage" aerial then could be pointed in any direction. It was an arrangement of crude but effective simplicity.

Although I have described the radar pulse transmitting aerial, there was, in fact, a second, identical aerial for receiving the signal bounced back by the target. The two aerials were set about 50 yards apart. The receiver aerial was rotated by the receiver operator, who watched the cathode-ray tube and waited for the aircraft blips to appear on the screen. For his part, the transmitter operator had to "bind" the transmitter aerial in such a way as to keep an indicator meter needle at zero. This indicated that the aerial was pointing in the same direction as the receiving aerial. Obviously, the transmitter signal had to be in the same direction as the receiver aerial in order for the receiver to detect weak pulse echoes from aircraft.

Once a transmitter was built, a radio receiver unit was needed that could work on the 1½-metre wavelength, which was considered very short at the time, and technicians worked flat out to build a suitable unit. A wide-band receiver was needed to capture the pulses, and what better equipment for this purpose than a television receiver, many of which were available in any radio store? Of the units available, the Pye 5 RF amplifier was chosen by Cockcroft for use in the new type of directional radar station. Soon after, the available Pye 5 RF units were commandeered for conversion to use as radar pulse amplifiers.

In a few days a prototype receiver was delivered to the Chain Home Low (CHL) Foreness Station, located in Kent on the mouth of the Thames Estuary, and was in operation with Cockcroft at the binder pedals. From start to finish he had solved the problem in under seven days, an astonishing accomplishment. Sited at the tip of the North Foreland, Kent, the beam could sweep north-east towards Germany and scan the danger area of the Thames Estuary.

It was now possible to track not only low-flying aircraft, but the magnetic mines as they descended by parachute into the water. Presented with a "mine map" each morning, the Royal Navy dispatched marksmen in small boats to explode the mines and thereby keep the estuary safe for shipping.

Cockcroft remained at the new station for some time to evaluate the performance of the new equipment, and to make improvements. For the next few weeks the CHL Foreness was the model from which RAF 60 Group produced other installations.

Another serious weakness of the CH system led to the Battle of Barking Creek, a rather costly affair in terms of aircraft damage. The main feature of the CH system was its long-range performance and its ability to read the height of an aircraft to within a thousand feet. Though height and range were fairly accurate (certainly enough for daytime interceptions), the system's bearing accuracy was not. This directional ambiguity sometimes led the operators to "read" an aircraft as being in front of the station when in fact it might be exactly the same distance and height behind the station—in other words, 180 degrees out of its supposed location. It was this feature that led to the battle.

A second contributory factor was the failure of fighter pilots to switch on the Identification Friend or Foe (IFF) transmitter unit that was mounted behind them in their aircraft. It is hard to understand now, but pilots, who had no knowledge of radar, did not appreciate the importance of switching on the unit before take-off. Alongside the "switch on" switch was another, wired to detonate a charge that would destroy the set to prevent its capture by the enemy. Many a pilot chose the wrong switch and blew up his IFF unit. The thud of a contained explosion, followed by the acrid smell of burning insulation in the cockpit, did not deter many pilots from destroying IFF units time and time again. Eventually the self-destruct switch was secured by thin wire to prevent its accidental use. It goes without saying that if the aircraft was without a functioning IFF unit, either because it hadn't been turned on or because it had been destroyed, the aircraft would be plotted by all radar stations as a potentially hostile intruder. Given these faults, conditions were ripe for the kind of debacle that occurred in the Battle of Barking Creek.

The battle occurred in the spring of 1940 during a visit by King George VI to Fighter Command Headquarters at Stanmore. Lord Dowding, A.O.C. (Air Officer Commanding), was explaining the working of the Filter Room when an aircraft that was not showing IFF was plotted by Canewden CH station. Since the station was on the Thames Estuary, the CH bearing showed the aircraft to be thirty miles out to sea or an equal distance in the opposite direction behind the station. If there was no IFF signal, the intruder, supposedly over the North Sea, had to be regarded as hostile, and on this occasion an "X-raid" was plotted by the filter officer.

Fighters were scrambled to intercept. Speeding out to sea, also transmitting no IFF signal, the investigating fighters were plotted by a second CH station and reported to Stanmore as intruders. A second X-raid was registered and more fighters were scrambled to intercept the first group of aircraft. The resulting confusion, in which British fighters fought British fighters, caused severe damage to a number of machines. The incident led to a tightening-up of the discipline in the use of IFF units. Even so, the Barking Creek affair was a costly lesson in aircraft interception, and did nothing for fighter-pilot confidence in the system being operated by Fighter Command Headquarters. There were a number of similar mishaps and X-raid intercepts of our own aircraft before the difficulty was finally overcome. The fact is that directional ambiguity of radar took a long time to sort out, which made life difficult during the Battle of Britain. Ghost tracks of non-existent aircraft led to the wastage of engine hours, wild-goose chases, and the exasperation and frustration of pilots. Despite these faults, the CH stations were the backbone of the radar system, and throughout the air war they performed well.

Because of its exposed position on the Suffolk coast, the Bawdsey establishment was moved to a safer location once war was declared. The first move was to Stonehaven on the Scottish coast south of Aberdeen, which was of course Watson-Watt's home territory. Here the institution became known as the Telecommunications Research Establishment. It was to be moved a second time to Worth-Matravers in Dorset; in early 1942, following a panic, it was relocated 100 miles inland at Malvern in Worcestershire, where, under the name Royal Radar Establishment, it is still located.

Following the short stay at Morecambe, we three "recruits" from Uxbridge were sent to the Radio School at Yatesbury, near Swindon in Wiltshire, where the confusion and lack of organization were soon obvious. No one knew who was in charge or what was supposed to be done, so that, for want of direction, and with no official RAF trade, we were at loose ends. On our own initiative, Bob Brown, Peter Friese-Green, and I organized a training programme and set up the first-ever school of radar. It is strange, looking back, that there were no ranks to distinguish the instructors from the instructed. There were a number of higher-rank non-commissioned officers (NCOs). Still, as far as we were concerned, we were simply civilians who happened to be in uniform.

For security purposes, the buildings allocated to radar instruction were some distance from the wireless operators' school. In hastily erected huts we uncrated the receiving equipment shipped from Cossor, the large manufacturer of radar equipment in Highbury Grove, and treated it as we would a new-born babe wrapped in cotton wool. The RF6 receiver was a monstrous piece of equipment enclosed in a wooden crate and delivered under the care of a military escort that had flanked the delivery truck on its journey from London to Yatesbury.

The RF6 was the very latest design, the acme of radar technology, incorporating every known device that scientific minds could devise. Imagine an eight-foot-high black box that, from drawing-board to delivery, had been subjected to every test and security check that the Ministry and the military could apply. In awe we uncovered the "black box" shrouded in white muslin, removed the cover, and stood back in admiration. But there, scrawled across the face of the equipment in white chalk, was a large hammer and sickle.

The moron responsible for this charming expression of political allegiance, presumably done during crating in Cossor's shipping department, had also vandalized the meter faces, two of which were damaged beyond repair. It was depressing. One must remember that, to the dismay of some but not many British Communists, Stalin had just concluded a pact of friendship with Germany. The question was how much of the new radar technology had already been given to the Russians by their espionage agents.

We instructed for long hours with little time off during this period over Christmas and New Year. I found it hard work teaching people who had only the most rudimentary knowledge of electrical theory. Lecturing, demonstrating the equipment, and checking the work results of students was an exhausting occupation; I felt relieved when the priorities changed and station-building was deemed of greater importance.

The Yatesbury experience was like a flash in the pan of my wartime experience, for in early February 1940 I was sent north to Scotland. Following Dr. John Cockcroft, I was to head a small team of technicians in installing a new station near the village of Rosehearty in Aberdeenshire. For me, the Rosehearty Station was the testing-ground of all I had learned at Bawdsey.

5

A SCOTTISH TESTING-GROUND

Rosehearty is a Scottish fishing village situated near a headland in Aberdeenshire known as Kinnairds Head. The village is about four miles west of Fraserburgh, a larger fishing community which, in 1940, boasted a small shipbuilding industry. Aberdeen lies sixty miles to the south.

With a group of technicians to help install the equipment at the new CHL station at Rosehearty (the buildings and compound had been constructed in a hurry), I reached Aberdeen by train. We made the last part of the journey by truck, rattling over a narrow road that snaked across the barren landscape. In summer, when the heather is in bloom, the country is covered in a rich purple carpet that is a delight to the eye; in winter, when gorse and heather are buried under snow, the terrain is forbidding and desolate. As we travelled north, the weather was freezing, and I felt as if the entire countryside was stretched tighter than a drum skin.

We were provided with billets in Rosehearty and were made welcome by the villagers in their comfortable grey-brown stone cottages clustered tightly around the tiny town square. Overlooking the village was the high hill on top of which the Air Ministry Experimental Station was located. With construction complete, our job was to make the equipment operational. In contrast to our own cozy quarters, a detachment of Gordon Highlanders assigned to guard the station had Spartan accommodation in draughty barrack huts in the compound, so I didn't envy them.

Kinnairds Head thrusts like an arrow into the North Sea. Jagged saw-toothed rocks fringing the shoreline form a dangerous obstacle to vessels moving along that coast. Many ships have been wrecked on the headland, and many mariners have perished on the rocks.

A glance at a map will show the strategic importance of Kinnairds Head in relation to the south-west coast of Norway, a fact that became obvious before the German invasion of Norway. In the spring of 1940, news of Germany's intentions came from the radio traffic monitored by the Enigma machine at Bletchley.

To put the existence of the Rosehearty CHL station in perspective, it is worth providing a sketch of the events leading up to the Battle of Britain and reviewing the Luftwaffe's opening offensive. Even before the Luftwaffe began its attempt to block the English Channel to Allied merchant shipping, and before passage through the Channel was made difficult by the German occupation of France, convoys were being routed around the northern coast of Scotland. This passage was dangerous enough without the added threat of air attacks from enemy bases in Europe.

Following the invasion of Norway, which began on April 5, 1940, Germany took control of the Stavanger Air Base on the south-west coast of the country. Stavanger lies 300 miles in a direct line from Kinnairds Head, so it was an ideal location from which to launch attacks on Allied convoys making passage along the east coast of Scotland.

The North Sea, and other waters off the coast of Europe, are divided into "sea areas" in much the same way that land is split into counties. The sea area of Cromarty, which lies off Kinnairds Head, was heavily mined. The shipping routes were narrow lanes swept clear of mines to permit the passage of convoys in line astern, which made them easy prey to enemy aircraft and U-boats. Construction of a radar station at Rosehearty was, for this reason, an urgent necessity.

The station was a modest enough installation. It consisted of two buildings for the radar installation and two barrack huts, one an orderly room and the other living-quarters for the guard of Gordon Highlanders. A 6-foot-high earth-and-sandbag wall had been built around the radar huts and the living-quarters to protect them from bomb blasts, and beyond the blast shield there was an encircling 10-foot-high barbed-wire fence. The ground between the outer fence and the blast shield, a zone about ten feet wide, was sown with high-explosive mines. The dominant feature of the site was two large rotating aerials mounted on the tiny huts. These were the same as the CHL Foreness aerials, since they were the same timber-hoarding-and-chicken-wire affairs, turned by the hand-cranked chains. The CHL had a range of about 60 miles and gave good coverage to the headland, so that we could detect and track aircraft at up to 15,000 feet, and plot convoys along the full 30-mile length of the Head shoreline.

We believed we had a superb early-warning system once the station was installed and operating, but the German position was stronger. They knew as well as we did that the convoy sea lanes along the coast were narrow—their U-boat captains had only to wait for ships to pass by in line astern—and the Luftwaffe at Stavanger was within easy striking distance. But this was not all. They had the Freya system in their favour, although we knew almost nothing about it at the time.

Freya was similar to our own CHL radar except that, in 1940 at least, it was better engineered than anything we had developed. Freya not only provided early warning of intruders, it rendered a clearer picture for the German operators because of the VHF feature of German equipment.

John Cockcroft joined us at the station and remained on site for a few days to supervise construction of the antennae and to check the equipment. Not being involved in the development of the first CHL radar, I needed instruction from John in the use of the new equipment before I could teach others, so I was grateful for his presence. He was unable to stay long because he was in demand at other sites, but I felt a lot more confident when I was left on my own because of the tuition he had given.

Training operators and technicians for the standard CH station was not difficult, but CHL equipment was physically more demanding. This is why, when CHL radar was in its infancy, we had to use all the experienced personnel we could find. We used Air Ministry staff (civilians) when we could get them because they were the people originally trained to operate the CH stations. The crew sent with me to Rosehearty was a mixed bag: two Air Ministry civilian technicians and three of us in the RAF. Additional RAF General Duties airmen arrived later.

Most CH operators, however, came from the ranks of the Women's Auxiliary Air Force (WAAFs) under the command of Chief Signals Officer (CSO) Dame Trefusis Forbes, usually referred to as the "Queen Bee". WAAF operators were trained mainly on operational stations until a school at Yatesbury was set up. They were introduced to the work as soon as they became available, but, until that time came, we dealt with the shortage of trained operators by working exhausting twelve-hour shifts.

The day John Cockcroft left us there was a heavy snowfall and a strong north-east wind that worked itself up into a gale. The hill was soon transformed from a merely bleak and disagreeable place into a snow mountain that made movement next to impossible. This was only the beginning, because, as the day wore on, more snow fell and was driven by the gale to submerge the huts to roof level. We came to feel that we might be buried alive.

Only the Gordon Highlanders who stood guard were unimpressed by the foul weather. Rifles slung from their shoulders, they patrolled the perimeter with stoic indifference, stamping their feet and blowing into their mittened hands in an attempt to keep their circulation going. Climbing up the hill for my tour of duty was a feat of endurance that required dogged determination. We decided to work around the clock because we couldn't trust ourselves to find our billets in the blizzard. The following morning we emerged into a frigid and silent world in which we seemed to be the only survivors, then we slipped and slithered down the ice-cream hill to our billets and bed.

According to my landlady, Mrs. Chrissie Buchan, the snow came in flakes "thick as porridge oats" while I slept. When I set off on the evening shift, it was mildly blustery in the village, but by the time I reached the station the wind sounded like the howling of banshees. We arrived to find the on-duty crew wrestling to steady one of the aerials in the gusts, while one of them struggled to repair a broken drive-chain. The crew that my crew was relieving hurried to sign the station log and depart as soon as we appeared, for they had had enough and were happy to let us finish the repairs. We spent a couple of hours making a patchwork repair, then went inside to finish the watch.

Not surprisingly, there was no air activity, nor was any expected, given the state of the weather. I hung on to the line to chat to the Stanmore operators, for the telephone link was a lifeline that helped make our isolated existence tolerable.

To write of making ourselves comfortable for the night is a relative term. There were no beds, so personnel who were not monitoring the screen or taking a turn at the transmitter aerial bedded down on the concrete floor behind the equipment. At that spot, warm air from the receiver power-pack circulated and made the concrete floor seem less Spartan. All the same, one rose for a spell at the equipment with a sigh of relief and a stiff neck.

There were two radar huts: one for the transmitter, the other to house the receiving equipment. Towards midnight during the second day of the storm, the transmitter operator telephoned to report that the drive-chain had snapped again and had snaked into the gear wheels. I went back with him to inspect the damage. It was obvious to me that repair in such weather was impossible, so we lashed the mast in a fixed position on Stavanger and hoped it would do.

Maintenance and repair of the equipment, but particularly of the rotating aerials, were a constant headache throughout the time I ran the station. The bicycle chains continually gave us trouble, and this first experience of breakdown during the snowstorm was frustrating to the limit of my patience.

Later, after the initial effort had been made to create an operational station, I had a chance to become better acquainted with our surroundings between snowstorms, and to appreciate the remoteness of Rosehearty and its distance from civilization as I knew it. It was certainly a lonely spot. At times, gazing out to sea or inland over the stark land, I felt as far from home as a Roman sentinel on Hadrian's Wall; nor could Roman sentries have been more uncomfortable than we were.

To work in the uninsulated wooden huts was worse than living in a meat refrigerator. The diesel electric generator unit was barely sufficient to operate the equipment, so the use of electric heaters was out of the question, even if we could find them, for there was no external power supply to the compound. We cursed fingers numbed with cold when making equipment connections, braced ourselves against biting winds when at work on the aerials.

I sometimes wondered if the sun ever shone on Kinnairds Head. For many weeks after we arrived at Rosehearty, ominous grey-black clouds rolled across the sky in never-ending succession when visibility was reasonable. Even in the foulest weather, cold or above freezing, the slate-grey rocks along the shore glistened like naked spears, and reminded one of the danger to ships that came too close to shore. Inland, the rugged hills were covered by a frozen mantle of snow and impoverished heather. It was easy enough to imagine Macbeth's witches croaking on a distant hill.

However, not all was gloom and doom. At the foot of the hill, the village lay snugly against the Highland winter.

At least we had comfortable billets there, unlike the guards, poor devils, quartered on the radar site. I don't mind admitting that it took a conscious effort to leave the billet and make the laborious climb up the hill for the next twelve-hour shift.

The Gordon Highlander guards were a clannish lot. They seemed to regard anyone from south of the border as a Londoner, whatever his home town in England might be. This probably had a lot to do with language, for we found the broad Aberdonian dialect, with its clipped words and phrases, almost impossible to understand. The station guards must have found our speech just as incomprehensible. There was also the problem of the disparity in our living-conditions, but despite these irritants we gradually reached an accommodation with one another.

To get to know them, I showed interest in armed combat and persuaded their NCO-in-charge, Lance-Corporal Forbes-Smith, to instruct me. He had served in the First World War, and was a master of bayonet fighting. He agreed to give me lessons once the weather improved, and he kept his word. As part of the instruction he frequently challenged me to combat as soon as I came into the compound for a tour of duty, and as long as he knew I was his inferior he treated me as the novice I was, ending the challenge with a mock thrust or jab of his rifle butt. It was all good, friendly instruction.

On the day I finally managed to catch him off balance, the Luftwaffe caught us napping. Neither experience was commendable, though I learned an important lesson, which was to stick to what I knew best. Somehow I had managed to get my rifle butt to Forbes-Smith's chest, which caused him to slip on the wet grass into a pool of mud. His Glengarry cap went flying, and his personality instantly underwent a profound change. Clambering to his feet, he charged at me like an enraged lion, thrusting, driving, parrying. He was breathing heavily and I prudently retreated, deflecting his thrusts, trying to keep my distance. But I was no match for the expert. In a sudden, lightning movement, he swung his rifle and struck me in the side of the face, sending me flying, spitting blood and teeth. Before the attack I had had two large front teeth. Now one was broken and the other loose. I pushed the loose tooth back into position and spat out the chunk of tooth into my hand. Forbes-Smith, rifle and bayonet at the ready, was standing over me, his face twisted in anger.

"You're mad!" I yelled. I thought he would strike me again, but he restrained himself.

"Lundoners!" he shouted. "Bluddy Lundoners! You come here throwing your weight about. Ye c'n drink our beer and rape our women, mon, but you'll never teach a Scotsman how to fecht!" He strode away in high dudgeon, leaving me wondering about his order of priorities.

Then came the shrill sound of the air-raid whistle and a shout of alarm from one of the technicians. I staggered inside, dropped my gear, and slumped on to the stool in front of the plotting-board. I took the head-and-breast set from the operator and spluttered "New track" into the microphone. Because of the delay, the intruder was nearer than it might otherwise have been—seconds are precious in radar tracking.

The operator, concentrating on the screen, didn't notice the damage to my face until he heard me speak in slurred tones. He turned and looked at me in alarm. I must have looked quite bloody. "My God!" he said. "What's happened to you?"

I explained and he laughed.

"I've told you a hundred times not to tangle with those mad Scotsmen. They'll chop you up for haggis! Here—give me the head-set and go and clean yourself up; you do look a mess. I don't understand you fellows one bit. You're supposed to be a boffin, so why worry about weapons and bayonet fighting?"

We gave Stanmore a running commentary on the visitor, a Junkers 88, until he was directly overhead, circling the compound. Then I stepped outside to observe him in the fading light of the calm and pleasant evening. Sunset came early there and the northern sky was aglow, suffused in crimson, with dappled pink clouds on the horizon. One of the sentries had turned out the guard with a bellowed "Turn oot!" and a half-dozen Highlanders under Forbes-Smith were now standing around with their rifles at the ready, waiting eagerly for the order to open fire.

I watched the Junkers 88 make another circuit and wondered what would happen next. One of the radar operators, manning a Lewis gun mounted on a tripod, was keeping the aircraft in his sights, itching for me to give the order to fire. He kept looking in my direction for a signal, but there was little chance, I thought, of our achieving anything, so I stayed him with a gesture.

The intruder orbited the station slowly to inspect us more closely. One well-placed bomb would have demolished the station completely. Soon the bandit was close enough for us to see the silhouettes of the pilot and the co-pilot outlined against the colourful sunset sky in the west.

The aircraft track, neatly plotted on the tracing-paper I later studied, showed that the Junkers 88 had made a beeline for the station. It was as though the navigator had used our radar as a homing beacon, which of course was not possible. Nevertheless, the Luftwaffe was obviously interested in our aerial gantries. Fortunately, I'd told the receiver operator to stop cranking some time earlier. There was no point in disclosing more than was necessary.

Showing contempt for us, the pilot switched on his navigation lights and came in for an even closer look, making the tightest circle yet. I thought he'd side-slip to the ground, but he didn't. It was a frightening experience all the same. The unsynchronized roar of the engines made a distinctive throbbing sound; the deepening sunset, and the strong, salty smell of the sea etched the scene indelibly on my consciousness. It was a cat-and-mouse game with the Junkers cat revolving slowly about our heads, ready to pounce, and we mice waiting apprehensively below. Then the pilot changed his tactics.

Flying directly over our heads, he banked steeply from the inland side and made a return approach as though on a bombing run; still no one made a move to scramble for safety. We watched, mesmerized, like wild animals caught in the headlights of an oncoming car. Coming at us head on, in a nose-down position, the black bomber had a menacing and ugly aspect. Then the pilot waggled his wings jauntily, switched off his lights, and flew straight out to sea.

As the throb of the engines died away, we were left standing in the hushed silence of the cool night. Nothing had happened. No bombs, no rat-a-tat-tat of bullets, nothing. The strange feeling of anticlimax was broken by one of the Highlanders who quipped, "He went awa' hame because he saw me affixin' ma bayonet."

During the relieved laughter that followed, I slipped into the operations room to report our not unfriendly visitor to the Filter Officer at Fighter Command Headquarters. I learned from him that the Luftwaffe had made similar inspections of a number of our radar stations over the past few days. Interesting!

At the time, I supposed it doubtful that the Junkers crew had ever seen a radar station, or even knew what function radar performed. If, however, our strange-looking installation with its rotating aerial was reported, and I had no reason to think it would be otherwise, Rosehearty would be marked for further attention, because anything new and different was naturally suspect. At that time the Germans had no specific knowledge of our use of radar. A few months later (as I discovered after the war), they photographed the CH and the two CHL stations at Foreness, using a telephoto lens from Calais. Later, during August, when Goering launched the Battle of Britain, there were to be attacks on radar installations on the south and east coasts, but the effectiveness of the overall system would not be jeopardized. The Luftwaffe did not seriously believe that we had advanced as far as we had.

A few days later our gentlemanly visit from the Junkers was followed by a less friendly approach in broad daylight. Fishing boats from Rosehearty, which the enemy might have regarded as our early-warning system, were machine-gunned, with heavy casualties among the fishermen. The villagers were incensed and looked to us to do something about the menace from the eastern sky, not understanding that we were as helpless as they.

The next week there was a daylight attack on Fraserburgh harbour, which caused heavy casualties, and then another on ships rounding the headland near the town. Two small vessels were sunk. The raids became more frequent and audacious, and it was not as though we did not "see" them coming. We identified each raid a good thirty minutes before the aircraft arrived, but once we passed our radar plots to Stanmore our responsibility ended, and we were powerless to take further action; CHL was simply a surveillance device then, not the eyes and ears of the fighter aircraft it later became.

Aircraft plots from monitoring stations such as Rosehearty were always transferred first to Stanmore. Stanmore repeated the information to Group Headquarters, who repeated to the nearest sector station, who finally issued orders to scramble its fighters for an intercept. It was a ponderously slow system of communication because by the time plots were registered, hostility was confirmed, and necessary instructions were issued, the raid was usually over and the bombers were heading for home. In the case of Kinnairds Head, more often than not the fighters from Dyce (Aberdeen), the nearest sector station, arrived to find an empty sky. It was a maddening business which left me wondering what we could do to overcome the problem.

No one knows the thought processes by which new ideas develop, and I find it difficult to provide a coherent and sequential explanation of how my system of ground control of fighters evolved. All I can say is that I thought about the problem deeply and expanded some ideas that were based partly on the navigational instruction I'd acquired before the war, in an effort to get a quicker response. I then wrote a report in which I proposed to short-circuit the communication system to permit ground stations to control fighters without disturbing Stanmore's overall control of Fighter Command. I sent the lengthy and detailed report to 60 Group HQ at Leighton Buzzard and waited for a reply. Without knowing it, I had proposed the Ground Control Interception (GCI) system more or less as it was eventually developed and put to use. When no reply came I was advised by Bawdsey contacts to forget about making improvements, and to concentrate on operating the station as it was designed to be operated. My job, they said, was to stick to operations, not to reorganize Fighter Command.

Although CHL Rosehearty was operating more than satisfactorily in its specific role, I was troubled that insufficient use was being made of the information we were supplying, so in the absence of an answer from HQ, I resolved to put my ideas of aircraft interception into practice somehow—illegally if necessary. With the help of William MacKenzie, one of the operators and an avowed Scottish Nationalist, or so he said, I experimented with a number of the suggestions I'd made in my report.

One of my ideas had to do with a combination short-wave radio we used in communication emergencies, a model TR 1082/1083 unit. Though old-fashioned, it was a sensitive set with which we could hear the Luftwaffe pilots on their aircraft sets as soon as they became airborne over Norway, and long before we could actually pick up their aircraft on our radar screens. Having passed our plots to Stanmore, we could watch the approach of enemy aircraft well before they made a landfall over the British coast. Their usual course was to sweep in a leisurely manner over the North Sea, machine-gunning likely-looking targets—fishing boats, convoys, and coastal shipping—before making landfall. Sometimes they turned north towards the Orkney and Shetland islands, sometimes south in the direction of Aberdeen. Nevertheless, on account of the position of Kinnairds Head, it was the Luftwaffe's most frequent and successful hunting-ground.

"Vector, eins, zwei, drei," came the guttural voices of pilots acknowledging course changes given by their leaders. The sound was unmistakable on my short-wave radio. What in fact was happening was that the experienced pilots, flying the Knickebein beam, were indicating to those under training the correct method of using the blind-bombing beam. I knew that they were employing some form of radio navigational system, without actually understanding what it was.

I first tried confusing them by playing tricks with what I took to be their navigational signals. By receiving the Stavanger signals, feeding them into our transmitter, and re-transmitting them, I was convinced that I was making matters more difficult for their navigators. On one occasion I tuned in on unmistakable German invective and persuaded myself that it had resulted from my handiwork, but at whom the angry words were actually directed I had no way of knowing.

I also tried jamming their homing beacon once they were on their way home, but I very much doubt whether I achieved anything worth while. Still, I was taking positive action, which was better than biting my fingernails. The methods I used to frustrate the enemy were hardly more than pinpricks, though they were interesting distractions and occupied the time on watch during which there were long periods of inactivity.

In the late spring of 1940, I visited the fighter sector station at Dyce for spare parts, and a chance call at the local pub resulted in a dramatic change in our interception technique. As a result, I was able to develop the ideas I'd put to HQ, except that my independence also landed me in serious trouble.

6

DECEPTION AND
INTERCEPTION

Time has faded the names, times, and precise sequence of developments, yet the important parts of the conspiracy remain as clear in my memory as though they had occurred yesterday.

Towards the end of May 1940, while the German breakthrough in France and the Low Countries was in progress, I drove into Aberdeen with MacKenzie to collect radio parts sent by the Cossor factory. Although MacKenzie was an operator without a thorough understanding of radar technology, he certainly grasped the essence of what I wanted to do to subvert the Luftwaffe navigational system, and was an enthusiastic supporter. He also shared to some degree my concern and frustration with the Rosehearty-Stanmore sector-communication system.

"Why not just contact the fighter pilots at Dyce?" he asked during the journey.

"That's the basic idea," I said, "but how? That's the question."

Direct contact was not easy. The set lines of communication were strictly controlled, and Stanmore jealously guarded its prerogatives.

We had a slow and uneventful drive through drizzly mist to Dyce and collected the parts we needed. We then dropped into a local pub for a pint before driving back to Rosehearty. In the pub, we got into discussion with some Czech pilots who were flying with the Dyce squadron. Their flight commander was a veteran from the Czechoslovakian Air Force, a volatile fellow who, in broken English, was holding forth on the lack of organization ("stupidity" was his word for it) of the RAF in "scrambling" the Dyce squadron time after time on a wild-goose chase.

"We come. The sky is empty. Luftwaffe have landed, and they drink the coffee while we drink the empty sky, yes?" He assured us that a thing like that would never have happened in Czechoslovakia, where, incidentally, he said, "ze beer is better."

As we were in uniform, joining in the discussion was easy. In passing, I said that we were radio-telephone mechanics stationed near Fraserburgh, which essentially was true. I said that from our equipment hut we often saw enemy aircraft flitting around the sea, and what if we were to pass the word to the Czechs? There would be ample time for them to intercept—provided, of course, we used the radio-telephone. It wasn't for us to interfere with the machinery of Fighter Command, I said, so perhaps the idea was not so good after all.

"Och! No, not possible," said MacKenzie in agreement.

"Despite the fact," I said, "that we have in our radio compound the forward-relay transmitter that controls their aircraft."

The flight commander and his fellow pilots pricked up their ears at this intelligence. In his own brand of English, the volatile commander said, "To hell with Fighter Command. What about it? You call us, yes?"

I explained the plan that had been forming at the back of my mind, and suggested we use some sort of code to let them know we were the source of the message. We said we'd pass the word on our radar plotting line to Schoolhill CH Station in Aberdeen. They, in turn, would pass the "scramble" message through their telephone tie-line connected to the Dyce sector operations room. In this way we would bypass Fighter Command and Fighter Group headquarters.

On receipt of our signal, the Czechs were to have a section of Hurricanes take off ostensibly on a "cross-country exercise", which would be reason enough to keep the aircraft free from the usual sector control. They were to fly to the Kinnairds Head district and we would give them directions on the radio-transmitter (R/T) as arranged. (We couldn't tell them that we would only pick them up on the radar as soon as they gained a reasonable height.) Having agreed on a simple code to identify villages and other main features of the headland, we were, in aircraft-control terms, in business. Whether or not the Czech commander realized that he was being set up I don't know, but it was an arrangement with which he enthusiastically agreed.

Back at Rosehearty I fixed a transmit/receive control for the VHF transmitter at the station, using a General Post Office Type 14-switch for the purpose. Then, using a field telephone borrowed from the Gordon Highlanders and connected to the transmitter modulation line, I could be in contact with the fighters overhead at the flip of the switch. We kept our messages short and sweet, which must have been puzzling to those who kept the radio-telephone log at Dyce (all radio traffic was logged). The Dyce station had operators on the same radio frequencies, so we could not prevent their listening in.

"Apples, angels three" meant "Enemy aircraft at 3000 feet in the vicinity of Fraserburgh". This information gave the fighters an accurate fix on the "bandits" at that moment and they would no longer have to rely on delayed information from Stanmore.

There were, of course, a few false starts. The take-off order, "Scramble", didn't get through to the Czechs as intended in the beginning, but our friends soon overcame the deficiency by planting a contact in the Operations Room to ensure that messages were properly communicated. For our part, we arranged that either MacKenzie or I would always be on duty during daylight hours.

The arrangement we made was soon working smoothly, for we caught a Heinkel III napping. He managed to wriggle out of the ambush, but at least the Hurricanes made contact and the Czechs were jubilant. A few days later a Heinkel was hit and crash-landed on the coast two miles west of Rosehearty. Then the Czechs brought down a Junkers 88, which ditched in the sea off Fraserburgh. The Junkers crew launched their dinghy, but their luck may have ended there because some of our Gordon Highlanders (not from the Rosehearty detachment) set out from the harbour in a rowboat to greet them. I didn't hear the end of this story.

As the efficiency of the system improved, the number of interceptions increased. Whether or not the Germans realized what was happening is hard to say. They may have had some inkling, because, within two weeks of the start of this system, the Luftwaffe delivered their first really damaging attack on Fraserburgh. The port could be legitimately regarded as a military target because a new type of magnetic minesweeper was under construction in the shipyard.

Evacuation of the British Expeditionary Force and remnants from French and Belgian units took place from Dunkirk in June 1940. Paradoxically, life became less hectic for us after that, mainly because of the influx of personnel to the station from amongst the ex-Dunkirk evacuees. The Dunkirk survivors had to be accommodated somewhere. Rosehearty got its share of additional guards, administrative personnel, and would-be operators; we had a surplus of both, so it was even possible to visit Fraserburgh and occasionally take a trip south. I was in the town's only cinema the night the damaging raid took place, and left the flickering screen to the sound of exploding bombs and air-raid sirens. There were no anti-aircraft batteries in the area, so the bombers had free rein in the night sky. The cinema-goers, having little conception of the horrors of war, stood about in groups watching the activity; nothing as exciting had occurred in Aberdeenshire in the past hundred years. The sight of the local fire brigade tackling fires caused by the bombs was a novelty, and a ludicrous one at that. Almost no one heeded the warning to get under cover; it was too sudden, too strange, too bewildering.

The "fun", if that is what one might call it, ended when a bomb hit a pub along the street and a section of a high wall fell on the gawking onlookers. The resulting chaos was terrible. First, the crowd dived for cover, then people hurried back to free those trapped under the wreckage. I joined in and we struggled frantically to remove the injured and the dead from the rubble, the burning timbers, and the choking dust. Even so, it took time for the enormity of the attack to sink in. An elderly man who was in the pub when it was hit kept repeating, "What a waste of gud whisky! What a waste of gud whisky!" as we carried him to a waiting ambulance.

The night ended badly for me. During my absence, a clerk at the station reported my private line to the transmitter in the radio-telephone hut to the military police. Things get around and someone had evidently commented on my system of aircraft interception. They correctly detected my secret operation, but attributed it to the wrong intention: the clerk reported the matter to the authorities and set a hunt in motion for my arrest for espionage. An armed escort was dispatched to Fraserburgh to find me and take me into custody. Fraserburgh was then a small enough place for them to locate me without difficulty and put me under arrest, and from that moment on I was treated as a spy.

A spy! It was ridiculous. I would have laughed, except that no one else thought it was a joke. Corporal Pierce, the NCO in charge of the arresting party, also suspected that I'd engineered the raid on Fraserburgh. Would I have been at a picture show in the targeted area if I had, I protested? The trouble was that our radar installation was still a closely guarded secret, and I was conditioned to say nothing. I came and went as I pleased, had contact with colleagues in far-away places that no one else knew about, and, because of my position, was not subject to the normal discipline of service life. It was not that I flung a cloak of mystery about my work, but rather that I kept my own counsel. The outsider always stands out like a sore thumb, and this could have led to resentment. Only the operators knew what function the radar equipment served, but not even they were aware of its inner mysteries, so even they couldn't or wouldn't vouch for me. The Germans, I'm sure, would have appreciated the irony of employing a Cockney Jew as a spy.

Back in Rosehearty I was allowed to sleep in my own billet for the night, with an escort present in Mrs. Buchan's living-room. The next morning there was an air of expectancy on the station when I went to collect some personal belongings, but no one was permitted to speak to me. I felt like a pariah, shunned by everyone, with not even a "Fine day" from the guards. About mid-morning a signal was received, instructing that I be taken to London under armed escort. Being held prisoner is a horrible experience, especially when you are uncertain of the charge. I was guarded from the time of my arrest until I reached my destination.

Carrying just my kitbag, I was taken by train to Aberdeen and from there transferred to an overnight express for the long journey south. Even my escort was caught up by the espionage charge, and stolidly refused to enter into conversation. His silence throughout the journey and the seclusion of the reserved compartment created so depressing an atmosphere that, by the time we reached the outskirts of London, I had convinced myself that I was headed for the Tower of London to be shot as a spy.

We arrived in the city during the small hours of the morning. London was dark and gloomy in the blackout and there was a long wait for the early workmen's train that would take us to Leighton Buzzard, headquarters of 60 Group. I was cold, miserable, tired, and hungry. At the Leighton Buzzard railway station, transport and additional guards were waiting for us. Still nothing was said to me, though suspicious glances and stony faces were offered by the military-police guard that met us. In normal circumstances a visit to 60 Group HQ, holiest of holy citadels of RAF radar, would be like making a pilgrimage to Mecca. As it was, I sat in the back of the three-ton truck gazing past the gun of the guard at the receding road. I was too miserable even to think about escape.

Accompanied by my faithful escort, I was marched across the yard to the front entrance of the HQ building, where an armed military policeman demanded our authority to enter. This produced, the welcoming party took its departure, while my guardian and I were directed inside to await the arrival of the Duty Officer. There was yet another long wait in a cavernous corridor dominated by a high ceiling and drab, greenish-brown bare walls. I certainly sympathize with prisoners who spend interminable hours waiting for persons in authority; nothing is more destructive to one's morale. My own, by then, was at its lowest ebb. I suppose my guardian was used to this sort of thing, because he remained unflinchingly silent. He must have had the patience of Job.

After what seemed an age, the Duty Officer arrived and directed us to another office, because, he said, he knew nothing about my case. Even in my demoralized state, it occurred to me that this was typical. We were then ushered into the presence of the officer-who-knew-about-the-case by a warrant officer carrying under his arm a large and bulky file. The file no doubt contained a list of my crimes against the national security. It would, I conjectured, include unauthorized modifications made to top-secret equipment; interference with VHF communications to fighter aircraft, thereby endangering the lives of pilots; unauthorized communications by a secret telephone line; and unlawful and treasonable communication with the enemy. These were but a few of the charges I contemplated as I stood before the flight lieutenant who was now studying my file.

I gazed at his face while he was reading the papers and experienced a growing sense of alarm, for his expression was stern and inflexible, the corners of his mouth curving grimly downward. He was hardly more than thirty years old. I was so prepared for the worst that his first question caught me completely off guard.

"Well! How did you go about it?" he asked. Before I had a chance to answer, his expression relaxed and he said, "Have you had any breakfast?"

"No, sir. We came straight from King's Cross and didn't stop to eat."

He addressed the warrant officer. "Please see if you can organize some tea for us."

"It's a bit early, sir."

"Then use your good offices with the tea swindle."

"Very good, sir."

The WO left the room and the officer spoke to the escort, who was still present. "Thank you for your services," he said pleasantly. "Would you please leave us alone now? This matter does not concern you."

The escort, too, said, "Very good, sir. If you would please sign this receipt for the prisoner."

The receipt was signed and he left. As soon as we were alone, the flight lieutenant got down to business. "Right!" he said. "Pull up a chair and let's see how you went about it."

With a huge feeling of relief I explained how, with a minimum of interference with sector control, I placed fighters in position to intercept intruders.

"Look!" he said, interrupting me, "I know nothing about these supposed contraventions"—he tossed the charge sheet aside—"I'm only interested in your proposal for the ground control of fighters, Nissenthal." He tapped the paper on his desk and, to my surprise, I recognized my own handwriting. It was the report I had written on fighter ground control. He grinned. "Let's forget about the stupid accusations."

The warrant officer brought in a tray of steaming dark-brown tea and placed it on the desk. The flight lieutenant stopped talking while the WO was in the room, and, except for saying, "Thank you. That will do nicely," he waited until we were once again alone before continuing.

"Since you've taken it upon yourself to deal with the problems of interception, how would you like to have a station doing just that kind of work?"

"Sounds wizard to me," I said, "but I'd like to know more about it."

He smiled again and offered me a mug of tea. "Milk? Sugar? Help yourself." I did. "We're planning a series of ground-control interception stations that will do precisely what you've described in this paper."

I was flattered, and wondered which had come first, the new system or my paper. Was there any point in asking? I was, after all, as low on the RAF totem-pole as it was possible to get and he was my superior officer. It wouldn't do to be too bold or inquisitive considering the fix I'd been in so short a time ago. I held my tongue.

"The first of these is under construction at Westhampnett near Tangmere," he said. "If you asked me, I'd say it's almost operational." He was wrong, as it turned out. The new stations were far from operational, which was fine by me, for I was again in on the ground floor of an important new development in radar. The prospect of working on the problem of combining radar and navigation opened up tremendous possibilities. I was overjoyed and itched to get on with the work, so I was rather disappointed to learn that I first had to return to Rosehearty to complete yet another modification to that installation. We seemed to be forever making changes. Still, I was excited to know that I'd be posted back to my parent unit, and to the Royal Aircraft Establishment at Farnborough.

The return trip to Scotland was interesting for a number of reasons, apart from the fact that I could enter the compound as a spy returned from the dead, and could cock a snook at those who had so easily condemned me. It meant that I was reinstated without benefit of a court-martial; I could say goodbye properly to my friends at Dyce Control and Rosehearty, and to my landlady; and I could let everyone know that I wasn't in absolute disgrace.

However, the really interesting experience that came out of my return visit was a discovery that altered forever the erroneous belief that radio waves travelled in straight lines. Conventional thinking was that radio waves, like light waves, did not bend with the curvature of the earth but soared off into space. The straight-line theory, to which I subscribed, was shattered one evening on radar watch.

I was intrigued by strong, even massive, signals coming to our receiver when the aerial was pointed in a north-east direction. I remember thinking it odd that I hadn't noticed them before. When the aerial was rotated, the signals vanished, so I had the binder crank the aerial back to the bearing of the blips and began checking to make sure that they weren't ghost signals. I studied them thoughtfully, consulted a map of the Scandinavian coastline, and finally came to the only conclusion that made sense to me. The strong response was nothing less than a long-distance echo off the Hardanger Mountains in Norway, over 300 miles away. This meant that the supposedly impossible was happening: our 1½-metre transmissions could be, and in fact were being, bent around the curvature of the earth, giving us a reflection off the Norwegian mountains.

It was an important discovery, which had profound implications for the future of radar. I was so excited by the revelation that I spent the rest of the night plotting the data we were getting and wrote a report. The next morning I sent my notes, calculations, maps, and tracings to Dr. Ed Bowen, who was setting up a second filter room (the first, of course, being Stanmore) at Kirkwall in the Orkney Islands. Ed called me as soon as he received the package and offered his congratulations. He was enthusiastic, for the discovery opened up all kinds of possibilities, especially in navigation. Among other things, because of the navigational angle, it meant that a new and more reliable aid to bombing would be available to the RAF. Bomber pilots were still navigating by the stars, as mariners had done since time immemorial.

I left Scotland with some misgivings. I would especially miss the comfort of my billet with Mrs. Buchan and her family at 37 Pitsligo Street. Rosehearty had been spared the violence of war, apart from the inconvenience caused when magnetic mines were washed ashore in the village, and there had been no bombing such as that experienced in Fraserburgh. The centre of the Buchans' rambling house was solid granite, built to withstand an earthquake. As I said goodbye, I warned Mrs. Buchan to dive for cover in this granite-built area if they so much as heard the sound of a bomber, and she promised to heed my advice.

Soon after I arrived in Farnborough, I learned that there had been a raid on the Rosehearty CHL station. The bombs missed and some fell on the village. Mr. Buchan wasn't at home at the time, but his wife and their daughters, Jean and Evelyn, were. They were found under a bed in the tiny granite section, unhurt, the rest of the house totally demolished about their heads.

7

RADAR AND THE BATTLE
OF BRITAIN

The Battle of Britain, which began some time after the collapse of continental Europe in June 1940, intensified throughout June, July, and August, and ended on September 15 in a spectacular way. The battle had no specific beginning, but we can be fairly certain about its end, even though air raids were a feature of wartime Britain long after the battle was over. The great air battle that raged over southern England was almost continuous in its final stages. Thousands of aircraft were involved as wave upon wave of bombers came into conflict with fighter squadrons of the RAF as well as with the numerous anti-aircraft batteries in and around the major cities.

Bombing night and day, dogfights during daylight hours in the sunny summer skies above London and the southeast counties of England, and sheer exhaustion on both sides, took toll of the fighter pilots and the air crews of both the RAF and the Luftwaffe. The battle was the greatest clash of men and machines in the sky that the world has ever known, and nothing like it has taken place since.

Following the fall of France, British prospects for resisting the Germans much longer were bleak. The army of the British Expeditionary Force, in its retreat and evacuation from Dunkirk, had lost its guns and its mobile equipment; the army was defeated and needed time to replenish its arms and equipment. On the plus side, the Royal Navy was strong in men and ships, though its strength was spread across the seven seas; and the RAF was intact.

Intent on invading Britain and securing his western flank, Hitler ordered his General Staff to prepare for crossing the English Channel. The invasion plan, which was to include assault by air and sea, was code-named "Sea Lion", but to assure success the Luftwaffe had to gain mastery of the skies. It was this need for mastery in the air that led to the Battle of Britain.

To prepare for Sea Lion, Germany had to do three things. First, it had to destroy Britain's south-coast and south-east-coast radar network, which by then Germany knew existed. Secondly, the Luftwaffe had to destroy the fighters of RAF Fighter Command. Thirdly, London had to be bombed into submission, to be reduced to a pile of rubble, thereby breaking the country's morale.

Despite the losses during the battle for France, the RAF had made a remarkable recovery. The radar warning network had been augmented by the building of many more CHL radars at strategic points and pairs of CHL radars were placed astride each CH installation. Also, new CH-type radars were being commissioned at a prodigious rate.

To appreciate how matters stood on the radar front in mid-1940, it is worth comparing British and German achievements again, for a number of changes had taken place.

Thanks to communication intelligence obtained from the Enigma intercepts, beginning on June 12, 1940, the deductions of Dr. R. V. Jones of Air Intelligence led to the discovery and analysis of the blind-bombing radio beams based on the Lorenz and Knickebein transmissions. As we now know, Dr. Hans Plendl of Germany developed the X and Y bombing beam systems before the war and then guided German radio scientists in the development of the Knickebein blind-bombing (long-range) system.

The Enigma provided other intelligence essential to the management and control of the Allied air-defence system, such as an estimate of the strength of German serviceable aircraft, operational and reserve pilots, and spares and fuel stocks available to the Luftwaffe in preparation for Operation Sea Lion, which included the aim of knocking out the British radar stations. Goering had issued the order for all Luftwaffe units to report their operational status, and, as a result, over a period of forty-eight hours Bletchley made a total count of 2400 fully operational aircraft—a frightening total. We had only 400 fighters with which to oppose the air armada.

Following my posting to the RAF Farnborough, and for the next three months, I was all over the place, working on equipment design and development, trouble-shooting, and dealing with radar-equipment operating problems. Though I was a foot soldier, so to speak, of the scientific community and never in a senior position, I was privy to learned secrets of German activities.

60 Group, under secret cover, always ensured that I personally received a top-secret fortnightly news sheet called "The RDF Bulletin". In addition, because I had been involved in radar developments from the beginning, I had easy access to the scientists with whom I had previously worked: Watson-Watt, Priest, Blumlein, and others. I therefore had an up-to-date knowledge of our own developments and made my contribution to them. Flying Officer Thomas Hunt was the person to whom I nominally reported at that time, and it was from him that I first learned about the magnetron, which was undoubtedly the most important invention in the entire radar story.

For the first time a single piece of metal was a complete radar transmitter. The cavity magnetron worked on the hitherto impossible short wavelength of one-tenth of a metre. Compared to what both the Germans and we had so far achieved, this was nothing short of miraculous. The shortest wavelength on which either side operated up to that time was ½ metre. The wavelength of the CHL system was

1½ metres and that of the Freya 2½ metres. The CH, of course, worked on the very long wavelength of 12 metres.

The Germans, the Americans, and the Dutch—in fact, everyone in the radio fraternity at that time—had decided that ½ metre was the shortest possible wavelength. For this reason, the magnetron, with its $^1/_{10}$-metre transmission, was a quantum leap in radar. Credit for its invention went to two young lecturers, J.T. Randall and H.A. Boot, from the University of Birmingham.

Randall and Boot considered the problem from a new theoretical standpoint, the essence of which was this: long-wave transmitters have tuning coils made up of many turns. Short-wave transmitters have coils with a dozen or so turns. Television transmitters, working on a very short wavelength, usually have only four turns of wire on their tuning coils. In other words, the fewer the coils, the shorter the wavelength. (It is desirable to have the shortest possible wavelength because it is easier to control its direction.) A one-turn "coil" (or what was in fact a very short, straight piece of metal called a "lecher bar") produced the shortest wavelength possible at that time. The limit of miniaturization attainable with current manufacturing capabilities had been reached.

But, following the logic further, the researchers asked themselves how they could get a shorter bar. It was then that they hit on the idea of taking a very short bar, drilling out the centre, and using the *inside surface* of the hole as the "coil"—and presto! the hole did resonate just like a coil. The "cavity magnetron" was born.

But the next step was the crowning touch. The engineers took a square slab of copper and drilled eight holes in a circular configuration with a further hole in dead centre. A slight crack from the inside edge of each of the eight holes led to the central hole where electrons were made to rotate at high speed. This constant circular movement of the electrons past the cracks made each of the smaller holes oscillate violently. For reasons that were fully understood only much later, the current was magnified exponentially by the adjoining coils. Thus, transmissions greater in strength by several magnitudes than anything yet achieved were possible. The coils became so hot, in fact, that the copper used up to that point simply melted after short usage. The magnetron proved to be the most significant scientific breakthrough in radar technology and allowed us to leap as far ahead of the Germans as their Knickebein outdistanced navigation by the dead-reckoning method that was still then being used by the RAF.

By experiment, trial and error, and the selection of various materials, Randall and Boot eventually settled on molybdenum (a high-melting-point silver-white metal that is brittle but easily oxidized by heat) as the best choice of material for the magnetron. In comparison with any other design of radio transmitter, such as the CH and CHL systems with their massive aerials, the magnetron was a miniature marvel; it could be held in the palm of the hand. This ingenious device set the stage for a new phase in the radar war, and gave the Allies a commanding lead, which they maintained until the war ended.

There were other exciting developments such as Ground Control Interception (GCI), in which I had played an initiating role. Airborne Interception (AI), which was closely related to GCI, was developed and improved over the next two years. In mid-1940, from the onslaught of the blitz and throughout the Battle of Britain, the AI system then in use (dubbed "smeller" by Churchill) allowed night fighters to approach to within 1200 feet of an enemy bomber. This distance was too great to be of practical value to the fighter pilot because the resolution of the image on the AI screen was lost at closer than 1200 feet. A method had to be found to bring the night fighter to within 500 feet of the target and to allow the pilot to identify the enemy aircraft visually, even if he only caught sight of the bomber's faintly glowing exhaust-pipes. Once it was located, the pilot could identify an aircraft by its silhouette.

The failure of the AI unit then in use contributed to the forced retirement of Air Chief Marshal Sir Hugh (later Lord) Dowding at the end of the Battle of Britain—a battle that some at the time accused him of losing, since had the AI been more efficient, more enemy bombers would have been knocked out of the sky by our night fighters in the closing months of 1940. Dowding, along with the Air Officer Commanding 11 Fighter Group, Keith Parke, was the victim of political intrigue. Yet, simply by preventing Sea Lion from taking place, Dowding won the Battle of Britain, though this fact was not acknowledged until much later. I'll discuss Operation Sea Lion shortly.

AI had originally been developed by Ed Bowen, but as he was fully occupied with other work in the Kirkwall Filter Room, Watson-Watt, who was in charge of improving the AI unit, contacted Alan Blumlein and put the problem to him. In a very short time Blumlein produced the "Blumlein modulator". Simply put, modulation means to change the form of a radio signal. The Blumlein modulator enabled a very short pulse to be obtained and helped to produce a sharp image on the radar screen at closer range, which enabled the night fighter to come within the necessary 500 feet of the target bomber on the darkest night. This was sufficient for the fighter pilot to complete his interception and to see the target.

The modulator, now in service in every radar installation around the world, is yet another Blumlein invention to add to his already impressive list: high-fidelity sound recording and playback, television camera tube electronics, and stereophonic sound. He still had one incomparable gift to give to the people of England before his fiery death midway through the war.

In comparison with these British radar developments, German radio scientists made no dramatic advances during the early war period. Germany was on the offensive and had a well-developed system for that purpose; radar was a defensive system. Radar equipment would be used in offensive operations later, but this was not the case in 1940. Germany had the Freya and Würzburg systems, true, but those systems would not be improved until German might was put on the defensive, which it was not, on any front, at that time. Innovations are born of necessity, as the British proved when their backs were to the wall. As far as Germany was concerned, radar was, for the moment, put on the back burner of its scientific stove.

Since I was more concerned with the operating aspect of radar than with laboratory research and development, a front-line job in which results were immediately obvious appealed most to my temperament. Visiting the south-coast stations, fixing and maintaining equipment, troubleshooting, I carried with me one of the few "3339 oscilloscope" test sets we then had. The scope was, and is, the electronic engineer's main test instrument, without which few electronic faults can be located. While I was doing a spell of work at CH Dover in early August, the main transformer of the instrument short-circuited and burned out.

Being only a short distance from London, and knowing the Cossor factory, I decided to take the instrument in for repair myself. Balancing the bulky scope on the pillion of my motorcycle, I set off for London. I took the coast road through Whitstable, roaring along on a clear and sunny afternoon with hardly a cloud in the sky. On the Foreness radar screen we had picked up enemy formations gathering over France for an attack, but I didn't know the size of the raid until I reached the outskirts of London. It was a huge air armada, and a heavy pall of black smoke over the city was evidence enough of its size. Near Chatham I could see that the cloud was centred over the industrial East End, and at times the smoke blacked out the setting sun. Tilbury Docks in the estuary were ablaze. Black bombers in the evening sky were swarming like locusts, weaving and turning at will. It was an unbelievable sight, in which the Luftwaffe had the sky to themselves.

I'd seen a couple of downed Heinkels in fields in the vicinity of Herne Bay, and a burnt-out Messerschmitt 109 on a Chatham tennis court a few yards from where a group of people had been playing, but what were three wrecks as compared with the horde of bombers and fighter escorts filling the sky?

Crossing the Thames through the Blackwall Tunnel, I reached Highbury Grove to find the Cossor factory closed. Now what? The streets were deserted except for patrolling air-raid wardens and a tin-hatted bobby here and there. I needed a bed for the night, so I drove around Marble Arch, through Shepherd's Bush, and went to my uncle Lew's flat in Ealing.

There was evidence of the blitz everywhere. Rubble, fallen masonry, sticks of broken furniture, washbasins, shattered slates, collapsed walls, and wrecked buildings with sheared-through upper floors indecently exposing their contents for all to see. Unattended fires were blazing out of control, because the fire brigades had their hands full dealing with large fires elsewhere, and the atmosphere was one of appalling, incendiary chaos. People with bandages or crutches limped along the blasted streets. Above it all, the puthering clouds of black smoke were being driven into the deeper blackness of the night by the heat from buildings, broken gas mains, and chemical fires. In the distance, at the Tilbury Docks where fires were raging, the conflagration was worse.

I parked the motorcycle outside Uncle Lew's flat and pressed the bell. There was no answer. Surprisingly, this suburb of London was relatively untouched, though there was a strong smell of burning in the air, and the same overall sense of destruction. Otherwise all was peaceful, still, and neatly civilized.

I plonked myself down on the steel box that housed the

scope and waited. Then, like an apparition, the building caretaker arrived from nowhere and opened the flat with his pass key.

Inside, I dumped the instrument in the hall and headed for the bedroom. There was a photograph on the bedroom chest of drawers: my two uncles in khaki surrounded by flags, in the First World War. They had won their war. Mine seemed lost. I sank on to the bed and slept with my clothes on.

I was awakened at nine o'clock that evening by my mother and my aunt, standing by the bed, saying, "Wakey, wakey! Dinner's ready. Get yourself cleaned up!" as though I'd never been away. Uncle Lew was in the Auxiliary Fire Service (AFS), and was still out fighting fires at Tilbury with the entire AFS. Every available fireman from as far away as Chelmsford in Essex had turned out to fight the biggest fire since the Great Fire of London. My brother Harold, still at school, was apparently on the street, trading hot bomb shrapnel with his chums. Mother and my aunt didn't think it in the least odd that they should find me in the flat, and we ate a hearty meal and exchanged news before I left. It wasn't necessary to tell them where I was working or where I was staying. They took my silence for granted. The interlude over, I said "Cheerio, then!", drove to the Cossor factory to get the transformer repaired, and rode back to Dover without incident to carry on the war.

Another memorable experience occurred during the London blitz when I met Quentin Reynolds, the legendary American war correspondent, who was an out-and-out supporter of the British cause. I had to get to London to pick up spares for a bomb-damaged radar station in Kent, and was without transport. Someone thought I might cadge a lift from a couple of news correspondents who were travelling to the city from Manston Aerodrome a short distance from where I was working. I got a lift to Manston and asked one of the journalists, an Englishman, for a lift, but he was reluctant to take me. Just when I was about to give up, his companion, a burly, good-natured American, turned up and saved the day.

"Sure, fella!" he said. "Glad to have you aboard."

Needing no second bidding, I slipped into the rear of the Humber Snipe, an official car bearing 11 Fighter Group markings, and sat alongside the American. On his other side he kept his hand on a large valise that clinked and clonked whenever its contents were shaken by the movement of the vehicle.

"That's a hell of a place to put a radio station," the Yank remarked when we were passing by the CH Whitstable station.

Being in the flight path of German bombers, the station was perfectly located for the job it had to do. I said nothing, but the English correspondent tartly commented that we hadn't expected France to collapse so easily and make the English south coast the front line.

"Yeah! I guess you're right! Behind barricades, Paris could have held the Jerries at bay for years." Our garrulous American friend dived into his bag and brought out a bottle of French cognac, which he uncorked and handed to the driver, saying, "Here, fella!"

The English journalist looked askance when the driver drank from the bottle and handed it back. I was getting to like this American. He had a free-and-easy style, so different from that of his English counterpart, who was intent on maintaining an air of superiority in the presence of the lower ranks. I smiled to myself. I was never a great drinker, but on this occasion I accepted the American's offer and took a swig from the bottle.

The American was a talkative fellow, confident and relaxed, and it was only necessary to put in a few ayes and nays to keep him going. The driver was more responsive, and increasingly agreeable as he helped to empty the bottle. All this time the front-seat passenger maintained a stony, disapproving silence, but he became increasingly agitated by the erratic weaving of the Humber.

At last we came to a jerky, barely controlled stop in front of the *Daily Telegraph* building in Fleet Street and emerged unscathed from the car. The English journalist went in immediately—to write a feature article on the horrors of drinking in wartime, no doubt. The driver went on his merry way, and I accepted the American's invitation to a meal in a nearby pub behind the *Daily Telegraph* building.

It was a good feast of roast beef and Yorkshire pudding, which I ate with a hearty appetite, because I hadn't eaten all day. My American friend, who by then I knew to be Quentin Reynolds, spent a great part of the meal hurrying from the table to the bar to speak to people he knew, and to the telephone to call contacts in the city. He was far too busy to remain still for more than a few moments at a time, and he gulped his food as though eating was a necessary but distasteful chore that delayed more exciting pursuits. As soon as the meal was over, he said goodbye, paid the bill, and hurried into the blackout in pursuit of a news lead—as he said, "for the folks back home, fella!" It was not the last I saw of him, for we met again on two occasions before the war ended.

When the victorious German army had reached the English Channel they confirmed, through photographs taken by powerful telephoto lenses from Calais, the existence of radar equipment on top of the cliffs of Dover. There were the four 360-foot masts and four 240-foot masts of Dover CH. On the same cliff-top site there were two sets of CHL, four gantries which were exactly the same in general appearance as their own Freya radar.

They had not discovered that we had radar during the 1939 Zeppelin reconnaisance, and now, with the day on which the air attacks on England were to start (codenamed Eagle Day) only weeks away, Martini's worst fears were rea lized. It was far too late to design and build equipment to counter the British radar. The only alternative was to seek out and destroy all the CH and CHL stations that they could find, by concentrated bombing attacks.

On August 12, attacks were made on the CH stations at Pevensey, Poling, Ventnor, Foreness, Dover, and Dungeness. Ventnor, on the Isle of Wight was put out of action when a flight of Junkers 88s, coming in low across the Channel, climbed over Luccombe Chine and delivered a pin-point bombing attack. We put the damaged station back in service by bringing in a Mobile Radio Unit (MRU) from Kidbrook, near London. All the same, it was a week before Ventnor could resume operation, and then only at a much reduced pick-up range.

There were eight main CH stations on the Kent foreland. Between them, these stations plotted most of the Luftwaffe air activity in occupied France, so Stanmore had up-to-the-minute information on enemy air movements throughout the battle. These stations only infrequently suffered more than superficial damage from enemy raids, as it was almost impossible to blow down a mast without a direct hit.

The Luftwaffe heavily attacked the fighter airfields: Hawkinge, Lympne, Manston, Biggin Hill, and the rest. The Biggin Hill Sector Control station in Kent, which had been in operation since 1936, was transferred to a fish shop in the nearby village of Keston, and, with the help of technicians from 60 Group, Biggin Hill was soon back in business. In the same way that it had dealt with forward fighter stations in Kent, the Luftwaffe devoted its attention to those fighter stations in the north and west London areas: Kenley, Croydon, Hornchurch, and North Weald.

Goering came to believe that the Luftwaffe had destroyed the RAF's ability to strike back, and, as far as Kent was concerned, he was very nearly correct. He did not, however, reckon with the resilience of the warning system or of the fighter reserves that could be brought into action from the north-of-London airstrips and from elsewhere in the country. He lightheartedly told his audience during a speech on

German radio in August that if the RAF bombed Berlin they could call him not by his title, Reichsmarschall Goering, but by the Jewish name "Herr Meyer".

Although London had been heavily bombed in July and early August, by August 15 the stage was set for an earnest blitzkrieg of the city. Fortunately for Fighter Command, the Luftwaffe got off on the wrong foot. The circumstances, as picked up on the British radar network and later confirmed during the war post-mortem, were these.

The bomber fleet, complete with fighter umbrella, assembled over France and flew in armada formation toward the English coast. En route, a belated weather report indicated that there was heavy cloud cover over the target. The fighters received cancellation instructions via their radio telephones, but the bomber fleet, relying on wireless telegraphy (Morse code), were left high in the air without an escort. Already enveloped in dense clouds, the bomber pilots were not aware that the fighters had turned back and blithely proceeded to the target.

We found out what was happening by monitoring the Luftwaffe radio control network.

On this day, August 15, the Luftwaffe put in its maximum effort of the war, flying 2000 sorties. At no other time did it succeed in getting such a large attacking air fleet into operation. They lost seventy-five aircraft and suffered many casualties among their air crews. The RAF, flying over home territory, lost forty aircraft. Many more were lost on the ground, however, because, being neatly parked in rows, they provided excellent targets. After that experience, orders were issued for all aircraft to be dispersed and properly camouflaged. There were blunders on both sides: the Luftwaffe through poor communications, the RAF for not having the foresight to disperse aircraft parked on the ground.

The air battles—and there were many—continued at a furious pace during the last two weeks of August; I witnessed many of them while making my patch-and-repair rounds of installations in the CH chain, damaged earlier. During those two weeks, most of 11 Group Sector Control stations were bombed out. The south-east section of the radar chain was also in bad shape. The battle became a slugging match in which the RAF and the Luftwaffe were the combatants, and their respective radar systems the seconds who maintained a watching brief at the ringside, being emotionally involved of course, and giving expert advice between bouts. Like heavyweight boxers, the contestants wondered how much longer the adversary would last.

Churchill took retaliatory action by dispatching a number of long-distance bombing missions to Berlin. The bombing probes that preceded the first really damaging attack on Berlin on September 7 were marred by navigational problems, for without bombing beams of the kind employed by the Luftwaffe, the RAF had to rely on astronavigation. Not until the 1942 advent of "Gee", another Bawdsey invention, did RAF navigation improve.

After the September 7 raid on Berlin, the Luftwaffe, always ready for a laugh, took Goering up on his radio-broadcast offer and called him Herr Meyer thereafter. Hitler was not so amused. He ordered the Luftwaffe to step up its bombing raids on London. This was done, but he had to give up his hope of dealing the final blow to RAF Fighter Command.

Evidence of a tactical shift on the part of the Luftwaffe came on September 15 when, before breakfast, the south-east-coast radar stations reported yet another unusually heavy build-up of aircraft "tracks" over the whole coastal area of northern France. In the past, we had had considerable difficulty in estimating with accuracy the number of aircraft in any one attacking formation. We could gauge the size comparatively, but not by the actual numbers, so, in an attempt to get an accurate estimate, scout aircraft were kept airborne continuously, to assist in estimating the Luftwaffe bomber strength. By September 15 most of the defects of the defence organization had been rectified, and Fighter Command HQ, placing more reliance on the radar net, waited patiently instead of committing aircraft to battle prematurely.

Watson-Watt's radar network was put to the supreme test on September 15. Making one or two obvious feints in an attempt to mislead, the mass formations of enemy bombers headed in the general direction of London. By carefully plotting their progress, the radar network provided Stanmore with minute-by-minute information that enabled the fighter squadrons to be economically committed one section at a time. The first carefully positioned sections of fighters made contact with the incoming air fleet high over Canterbury and tore vertically through the tightly packed formations. Damaged German bombers, no longer able to stay in the protective embrace of the fighters, lost height and were pounced upon by low-flying Hurricanes and other fighters waiting below.

The surviving bombers closed ranks and flew on, only to meet fresh defending fighters, which, having remained aloft with a minimum of waiting time, descended to refuel and rearm in the knowledge that other fighter squadrons would deal with the bomber stream further along the flight path. The economic use of fighter strength based on the maximum use of the radar network constituted a radar-controlled ambush that had never before been achieved. Despite the onslaught of controlled fighter strength, the Luftwaffe persisted in the attack and suffered accordingly.

Air Vice-Marshal Leigh Mallory, Air Officer Commanding 12 (Fighter) Group, had for some time advocated "wing" attacks on bomber formations. (A wing is a combination of five or six squadrons flying as one unit.) Parke, commanding 11 Group, resisted Mallory's advice to send in wing formations and stuck to his policy of uncoordinated use of fighters in a fight of attrition. Once the enemy reached the target area, Mallory took over and used the fighters of his group in wing attacks. The two tactical techniques were complementary, for the completely fresh fighters from the north, operating in wings, were the answer to bombers swarming over London.

The Duxford Wing (five squadrons under the command of Wing Commander Douglas Bader), requested by Parke, arrived at the psychological moment, and the greatest dog-fight of the war took place in broad daylight in the clear September sky. Unloading their bombs, enemy bombers began the return journey, only to run into the refuelled fighters of 11 Group, committed in the same manner that characterized Parke's tactics during the approach journey. In the afternoon, the same bombers and crews repeated the morning's operations, to run the same gauntlet of fighters and meet the same reception over London.

September 15, 1940, marked the second climax of the air fighting and the last of the great day battles over Britain. On this day the Luftwaffe made 1700 sorties and lost 50 bombers, as compared with 25 fighters of RAF 11 and 12 Groups.

Without mastery of the air, Hitler decided to cancel the invasion of Britain, Operation Sea Lion—for that year at least. Fighter Command and radar had triumphed and the Battle of Britain had been won by the British.

8

THE TIZARD MISSION TO OTTAWA AND WASHINGTON

September 1940 was but one of the many turning-points in the ebb and flow of the tide of conflict in the Second World War. The air raids over London and other major industrial centres would continue, but the main point of the Battle of Britain, which was Operation Sea Lion, had been denied to the foe.

September also brought a change in my own fortunes, because from that time on, and for the next two years, I became immersed in the development of the Ground Control Interception (GCI) system. I was therefore not involved in one event in the broader development of the radar war that was then taking place. The event to which I refer was the Tizard Mission to Canada and the United States. The delegation of scientist-soldiers and top civil servants was so important a part of the radar-war story that it is worth writing of it here so that it may be included in its chronological order.

Britain desperately needed American help. Churchill knew that Great Britain could not survive without American support. I am well aware that the subject of Churchill's relations with President Franklin D. Roosevelt is a matter of documented record, yet the significance of the Tizard Mission has never been satisfactorily acknowledged. In fact, it was extremely important to Churchill's designs, as it served as an inducement to the United States to give material aid to Britain.

Henry Tizard, then, was entrusted with a commission vital to his country's survival. The secrets he carried were gifts of friendship that said, in effect, "The good citizens of Great Britain and Northern Ireland, and of the British Commonwealth, send you these gifts. Be our allies in our hour of need, and do not forsake us. It is our lives and our freedom that we are risking, but what we are doing is for the general good of all free people, including, in the end, the United States of America. We have the will to fight, but we need your help. Be the friends, therefore, that we think you are." The message was clear and unequivocal.

The situation for Britain was still grave. By the autumn of 1940, after one year of war, Germany had control of the whole western seaboard of Europe and Scandinavia, from the northern tip of Norway to Spain. Only the English Channel separated Britain from Hitler's Europe. Survival depended on Britain's ability to maintain a sufficient supply of food, munitions, and supplies to continue the fight. Tizard's job was to secure that supply.

The Mission, which included Bowen and Priest, sailed for Halifax, Nova Scotia, on the *Duchess of Richmond* on August 31 and arrived in mid-September. The Canadian prime minister, Mackenzie King, co-operated whole-heartedly with the distinguished visitors. Special security was arranged for protection of the cargo, and a heavily guarded train was laid on to move the visitors. Because Canada had already promised to use its available resources to help, and because there were to be discussions with Canadian experts, the train was routed to Washington via Toronto, Ontario.

In addition to radar equipment, the secret cargo included a number of interesting devices. One was a new "gyro gunsight", which was being fitted to all Spitfire and Hurricane aircraft and would improve American fighters. For use in high-explosive shells, there was a proximity fuse that detonated the shell in close proximity to the target; hence the name by which it was known. The fuse was based on a German design provided in the Oslo Report discussed in Chapter 4. Another was a partially developed gun-laying predictor designed by Colonel Derringer of the British Admiralty Research Establishment. The predictor, a height and range calculator that would today be called a computer, was to enable the Swedish Bofors gun to be used for anti-aircraft work. It was, however, the cavity magnetron, with which the British hoped to revolutionize warfare and defeat the U-boats, that was the crowning glory of the cargo of secrets being carried.

The gun-laying predictor, on which Colonel Derringer had begun work, was taken over by Dr. C. J. Mackenzie of the National Research Council of Canada as a project to be solved with urgency. (The Luftwaffe air raids on England were at their height, and the anti-aircraft guns then in use were inaccurate.) The theory behind the range and height predictor was sound, but the incomplete prototype needed a lot more work before it could be perfected. Mackenzie, in turn, handed the problem over to Research Enterprises, a Crown corporation in Toronto managed by Colonel Eric Philips.

In Washington, the scientists described the new devices to Roosevelt and an assembly of American defence experts, and revealed details of Britain's radar network. Priest and Bowen explained the warning system in detail, and gave what information was known about German radar, though this was sketchy. The American radio communication experts understood the significance of the radar network but, according to Priest and Bowen, were frankly sceptical of the claims made for the magnetron. For years they had been trying to develop a transmitter that would operate on a microwave length but had abandoned the project as unattainable.

Now the British scientists were claiming success with a lump of molybdenum that could be held in the palm of the hand, and yet emit high-powered microwaves at one-tenth of a metre. Dr. Mervyn Kelly, a leading electronics physicist present at the talks, was asked by the President to verify the British claims by testing the magnetron (in the care of Bowen) in his own laboratory. Within twenty-four hours he told the President that the British device, operating at low power, produced a thousand times more power than the most powerful American device, the "Klystron", was capable of doing. The British claims, he said, were fully justified by the results.

Despite the scientific knowledge that the United States received in return for the generous help it gave to the British, the U.S. Air Force was not quick to make full use of the information received. Far more than the magnetron was handed over. The entire radar warning system was explained, and important devices such as an Identification Friend or Foe (IFF) unit were delivered to the U.S. defence chiefs. There were, perhaps, production delays that prevented America's building a similar radar network to provide a watch over its territory. But, whatever the reason, the failure of the Americans to take immediate advantage of the British revelations led in part to the destruction at Pearl Harbor on December 7 the following year.

On that day an American radar team, using a network very much like the British CHL system, plotted the incoming Japanese air fleet approaching Pearl Harbor, but the IFF system was not yet then in use in American aircraft. As a consequence, American radar operators had nothing with which to identify the incoming aircraft as hostile. After Pearl Harbor, in an effort to make up for what had been lost by their lack of preparedness, the United States government requested, and obtained, the services of Watson-Watt.

The scientist was received in America with open arms, and treated with a respect and deference that is given in the United States as nowhere else to men of science and learning. Faced with a threat to its national security, the American defence chiefs acted with dispatch to make good the holes in the fabric of the warning network. Watson-Watt was listened to and his suggestions on the measures he thought should be taken were heeded. It is hardly necessary to say that there was no investigative sub-committee to ponder and weigh the advice given by the scientist.

Tizard and his team had done their work well. The Americans were convinced, Churchill was assured of increased U.S. support for the war effort, and Canadian participation in the necessary research, development, and production was arranged by on-the-spot consultation. As far as further radar research and development was concerned, Canadian participation was as important as the United States involvement. In the face of wartime production problems, basic research had had to take a back seat in Britain. Research and production facilities in Canada, particularly in the leading universities, were as good as any in the world in the early 1940s. There was a reason for this.

Canada already had a solid international reputation in scientific research in such fields as radio technology, physics, medicine, and organic and inorganic chemistry. The universities of McGill, Toronto, Queen's, Alberta, and Manitoba had well-funded research programmes. The National Research Council (NRC) of Canada, which had recently been formed, offered its research assistance. The second important factor that led to Canadian involvement in the radar war was Tizard himself.

He had strong connections in Canada through his sea-captain father, who, at the turn of the century, had mapped a large area of the Canadian north. Tizard Harbour was named in honour of Captain Tizard for his work in the exploration of northern Canadian waters. For this reason, and because of his own scientific contacts, Henry Tizard was well known in the country, and Canadian response to the request for help was immediate and well organized.

Dr. C. J. Mackenzie, Toronto chief of the NRC, was appointed to co-ordinate the national research effort, which included the research branch of the Royal Canadian Navy, the NRC, a number of universities, and private enterprise.

One of the most vexing radar problems concerned the difficulty experienced with gun-laying radar equipment. The trajectory of a shell could be calculated by any trained gunnery officer. The science of trajectory mathematics for ground-to-ground missiles was well known, but how far to fire a shell in front of a moving target such as a high, fast-moving aircraft was not so well established. There were added variables in the equation to be resolved, such as the elevation and speed of the moving aircraft, and the direction and velocity of cross-winds. In addition to muzzle velocities, shell-trajectory characteristics, and gun elevations, these factors had to be taken into account.

With sixty million dollars' worth of orders—no small backlog of work in 1940—Research Enterprises was a busy organization. Yet, Research Enterprises had its first microwave gun-laying radar, the GL.IIC, ready for service within a year of starting work on the project. The GL.IIC, of course, was of little use without suitable shells to fire, which is where Mackenzie's co-ordinating work with the various research groups was of value. A new proximity fuse, an improvement on the design brought from Britain, was developed by Dr. Arnold Pitt of the University of Toronto. The proximity fuse and the new gun-laying radar proved to be the principal means of defeating the German V1 (an unmanned flying bomb nicknamed the "Doodlebug"—from the droning sound of its engine). The Germans were to use the V1 against Britain in large numbers towards the end of the war. The GL.IIC and proximity fuse combination was largely responsible for the destruction of an estimated eighty per cent of all V1s.

The United States, with Bowen in charge of research at the Massachusetts Institute of Technology, also developed an excellent gun-laying radar based on information provided by Colonel Derringer through the Tizard Mission. This was the SCR 584, a lighter and more mobile unit than the GL.IIC. Unfortunately, the SCR 584 did not arrive in Britain in sufficient quantity to be used against the V1s. Because of its mobility, the SCR 584 was, however, used in the Normandy invasion.

Other radar and radar-related research was conducted elsewhere in Canada. In Halifax, for example, two University of Alberta researchers, Dr. R. W. Boyles and Dr. G. S. Field, who had carried out sonic research before the war, developed high-performance sonar equipment for the war against U-boats. Sonar, like radar, but using sound waves rather than radio waves, needed a special crystal to operate. Once the crystal research was complete, the system was handed over to Research Enterprises in Toronto to manufacture, and was then installed on all warships that could be routed through the port of Halifax.

Canadian research facilities were used in many other ways for the development of new ideas and devices. Because they are not a part of the early radar story, these will have to be told in detail elsewhere, but some are worth mentioning because of the people involved. For instance, Dr. W. Banting of the University of Toronto, working under the direction of his father, Sir Frederick Banting (co-discoverer of insulin), tackled the problem of blackouts suffered by pilots doing steep dives and tight turns during aerial combat; these were caused by the force of gravity and resulting loss of blood to the brain. He found that if the pilot's flying suit was padded with flat packages of water, blood was retained in the brain and blackouts did not occur. Banting's suit, which was to be introduced into service towards the end of 1942, was the forerunner of the astronaut's spacesuit.

Another Canadian contribution was research to improve the quality of British explosives, which was poor. Size for size, German bombs were twice as destructive as British bombs. The British knew of RDX (a powerful explosive), but because of its instability they were unable to use it for production purposes. Dr. George Wright of the University of

Toronto set up an "explosive" test bench in an unused elevator shaft in the university, a stone's throw from the government buildings of the Province of Ontario, and conducted experiments on detonation. While Wright worked out detonating techniques, Dr. John Ross of McGill University in Montreal developed the chemical discipline that would permit armament factories to handle RDX safely and in large quantities. Both Wright and Ross were successful.

Lastly, in the category of Canadian wartime research and development, there is the work of Lieutenant Neville Goodeve of Winnipeg, who joined the Royal Navy at the outbreak of the war. He initiated a system to protect ships from magnetic mines. In Halifax he developed the Double "L" sweep, a device that consisted of two floating cables towed by a minesweeper. When an electric current was discharged between the cables, magnetic mines in the vicinity exploded.

The Germans, finding the magnetic mines less effective, introduced acoustic mines, which were triggered by the vibrations of passing ships. Once again, Royal Canadian Navy researchers in Halifax investigated the problem and provided a solution.

I have sketched these weapon and scientific research projects to indicate the extent to which Canada put its research facilities to use, even though many of them have little to do with the radar war. The total research effort was enormous, and without question the efforts of Tizard and his team of scientists brought a rich harvest to aid the Allied cause.

9

GHOST STATION

Towards the end of September 1940, I joined the Air Ministry Experimental Station at Westhampnett, near Chichester in Sussex, the Fighter Interceptor Unit (FIU) of the RAF. Westhampnett was to be the first ground-control-interceptor station, specifically for the control of night fighters. In the same way that we at Rosehearty learned to work directly with the Czech pilots, the Westhampnett radar controller would direct the fighter pilots independently of the sector-control officer.

I was given a friendly though disinterested reception on arrival, and it did not take long to discover why. Under "Boffin" Brown, the project director, there was always a large contingent of Air Ministry engineers in attendance. Like most researchers not subject to service discipline, their dress and appearance were informal. Working long hours with those of us in uniform, they tested, modified, and experimented with the equipment until they were exhausted. Then they kipped down in the operations room, which did nothing for their appearance. Once refreshed, they got up, ate, and went on with their work. To the regular, non-technical service personnel, the boffins were a race apart, so that, to the orderly-room staff warned of my impending arrival, I suppose I was just another egghead with whom they had to deal.

The Westhampnett radar screen, cluttered with images of hills, buildings, pylons, and other permanent echoes from the surrounding countryside, was quite unlike the normal picture viewed on CH and CHL radars. The basic characteristic of the conventional high CHL aerial was to "see low", but over the sea. Over land every bump and knob of the surrounding country appeared on the screen. Although the surrounding topography was accurately displayed on the screen radio map, it was difficult to track and control individual aircraft because they were masked by the high ground. It was a case of not being able to see the wood for the trees.

I knew we had to get rid of ground images, to provide a clean screen, but how to do this was not obvious to me. The aerial was mounted on a frame at ground level. I put the problem to Don Priest, then at the Telecommunications Research Centre, and he suggested a simple remedy. By cannibalizing a GL.IIc gun-laying radar unit (the Canadians were at that time working on the GL.IIIc version), and mounting its aerial on one of the mobile cabins clear of ground level, we had the perfect answer.

The effective height of the rotating hoarding of dipoles was now ten feet above the ground. The modified GL aerial provided a sharper image on the screen, and cut out the interference caused by high ground and buildings in the vicinity even though operated only a few feet above ground level. The modification removed the clutter, but the equipment's operating range was reduced, a shortcoming that, for the time being, had to be accepted.

Apart from this defect, Airborne Interception (AI) radar used in conjunction with the GCI unit work had begun, and a number of successful interceptions were made by the new combination. There were, of course, difficulties to overcome, and a major one was the aircraft height-finding system.

Calculation of the target's height depended on a so-called "calibrated differential gain control" for the receiver. Differential gain control was a clumsy and inaccurate way of calculating height, but until a more reliable method could be found it had to do. Watson-Watt found a solution based on his own idea of splitting the two main receiving lobes in the vertical plane (a crib from CH radar). The vertical-split idea was used throughout the blitz in 1941. Until the magnetron was sufficiently developed for use in all height-finding operations later in the war, the vertical split was our only means of calculating height.

From the operational standpoint, individual aircraft range and bearing plotting, as practised both by the Germans and by us, was useless for high-speed fighter control, because it was too slow. So, to assist the operations staff, one of Watson-Watt's pre-war brainchildren, the Plan Position Indicator (PPI), was used. This was an electronic map superimposed on the radar screen. That is, an actual map of the station and the surrounding district to a radius of 60 miles was drawn on the screen. All aircraft moving in the air space of the area could be watched and plotted with ease. This system, still called PPI, is in use throughout the aviation and navigation world today—even for busy sea lanes such as the English Channel, the St. Lawrence Seaway, and New York harbour.

Another device already in use was adapted to GCI radar. This was the Craig computer. Colin Craig, an Air Ministry civilian who had taken the standard pilot's course, converted a standard course and speed calculator to work in reverse. The calculator, strapped to a pilot's knee during flight, enabled him to compute a course on which to steer. Craig's device permitted ground controllers to compute a complete set of corrected instrument readings for a given flight. These were the same readings that would be seen by the pilot, except that the "ground" readings were more accurate, because ground operators had fixed points of reference. This gave GCI controllers an advantage over the pilots of enemy aircraft.

On the basis of the research done on the Westhampnett Fighter Interceptor Unit, the first six mobile GCI stations were constructed. It was a scramble to train operators in time for the night battles that would begin once the "bombing season" opened. In January and February 1941, when these preparations were being made, the weather was truly awful, and flying next to impossible, given the lack of sophisticated flying aids. But it was taken for granted that, any time from March 1941 on, another massive assault on our airfields and cities would begin.

Each mobile GCI unit consisted of a transmitter vehicle, an aerial trailer, an operation-room unit, a diesel generator, and various miscellaneous vehicles such as a workshop. A complete station, either en route to a site or set up for business, looked impressive, yet the original units had shortcomings when it came to fulfilling their mission. In the first place there was the height-reading problem. In the second, there was still the distance problem of the AI equipment, for which Blumlein would produce his modulator.

Two of the units assembled were to be located in the West Country, and I was involved in both. The first one, still experimental, was located at Avebury in Wiltshire. It was built with the help of FIU and Yatesbury radar school technicians, and was for radar surveillance of the Bristol area. Avebury lies about thirty miles east of Bristol in a direct line. Apart from its importance as an Atlantic seaport, Bristol was the home of the Bristol Aviation Company, at which the new, all-metal night fighter, the Beaufighter, was being built. We needed a metal-built aircraft for night-fighter work because, of course, fabric-type aircraft lacked a proper metal "sounding-board" for generating radar pulses clearly. The all-metal Beaufighter proved to be a stable machine, and ideal for the use to which it was put.

Moving the convoy to Avebury was a saga I should not like to repeat. The convoy was almost half a mile long when stretched out, and getting it through the narrow roadways was like feeding a pig to a snake. The two aerials, one each for the transmitter and the receiver, were 10 metres long by 6 metres high, mounted on a cubicle GL.II trailer. The contraption needed an operator at the binding pedals as well as an observer, to manipulate the turning gear in order to negotiate sharp bends. Other units in the mobile column were almost as difficult to handle. The convoy filled the small roads along which it moved, and, because of obstructions such as low railway arches and narrow bridges, we frequently had to look for an alternative route and then shunt the entire column back. As much a problem as any was the convoy crew's too-ready willingness to decamp into the nearest pub whenever we were delayed by some obstruction. I was not enough of a disciplinarian to maintain order. As a result, the journey of under 200 miles took two weeks, so I was pleased to reach our destination and to return to Westhampnett.

A short time after completing the delivery to Avebury, I was instructed to report to 60 Group HQ at Leighton Buzzard for my next assignment. From Leighton Buzzard I was to be sent to 10 Fighter Group, the most westerly group of Fighter Command. At 10 Group HQ at Exeter I was briefed by the Group Sector Operations officer and handed a map on which was marked my next destination. It was getting more and more like a treasure hunt: "Report here. Go there. Get further orders. No, I don't know what it's about, Flight. You'll find out." Marked on the map in red was an RAF station called RAF Soar Mill. The location (a six-figure map reference was supplied) was Bolt Head, a headland on the south coast of Devon.

"Is there an airfield there?" I asked the briefing officer.

"No! But there will be after you get there," he said. I didn't know what to make of this conundrum and he couldn't explain. "Do your best, Flight. We'll send a mobile unit and personnel as soon as they're available."

"Very good, sir," I replied, mystified.

Armed, then, only with the map and carrying everything I owned in my RAF-blue backpack and canvas kitbag, I took an early-morning train to a little place called Kingsbridge. It was late February and the weather was terrible, but nothing could dampen my spirits. The prospect of operating an independent GCI station shone like a comet in the firmament of my imagination.

From Kingsbridge railway station I made my way to the hamlet of Malborough, near Hope Cove, and called in at the tiny post office to inquire as to the location of Soar Mill Farm.

"We don't want you military down here," the aged and sinewy postmistress told me. "We've had no bombs and we don't want none."

"I only want to know the whereabouts of Soar Mill Farm."

"There be no such place, so don't waste our time," she said snippily. "We've a lot of work to do."

I needed information, and the post office was the place to get it. "Right!" I said. "Where will I find the village constable?"

"I don't know. He'll be around here somewhere, maybe."

A slip of a girl behind the counter, the postmistress's granddaughter perhaps, piped up. "He lives in the cottage at the end of the street."

Looking cross and testy, the postmistress gave the girl a withering glare.

"Thanks," I said and left, but in taking my departure I noticed the sharp-tongued postmistress dive for the telephone. Perhaps that explained why, when I reached the cottage, the policeman's wife said he was not at home and she had no idea where he might be. Malborough village had one street, so it was hard to imagine the only arm of the law in the vicinity getting lost doing his duty. All the same, it was a rum business coming to this little place where the natives seemed unusually hostile.

Using my map, I spent the morning tramping the Start Point region of Bolt Head, looking for the spot called Soar

Mill Farm. I wandered for an hour or more over the large headland that juts out from the Devon south coast and, with the aid of map and compass, convinced myself that I was on the spot where RAF Station Soar Mill should have been. Instead, all I could see was a sea of uncut grass strewn with the pancake droppings of cows. Although they were on the headland, with high cliffs that fell precipitously to the sea, the fields of Soar Mill formed a shallow, saucer-like depression, dewy green and lush under the cloudy Devon sky.

In the distance, across Salcombe Estuary, I could see the radar tower of CH West Prawle, which was one of the first stations of the radar network to be built. Convinced that I had got the right place, I trudged back to Malborough post office and prevailed on the postmistress to get me a trunk call to RAF Station Exeter. Guardedly I explained my predicament to a disinterested orderly-room clerk. In the languid tones of one who is used to dealing with crank calls, he told me I was loony—as if an RAF fighter station could get lost! He then disconnected the line.

This was my introduction to 10 Fighter Group. I had been posted to a non-existent station that someone, somewhere, had obviously marked on a map in a place that they thought was a good location for a station. There was worse to come.

As I was about to leave the post office for the second time, the young girl told me with great excitement that a contingent of RAF boys had arrived at the Kingsbridge railway station and was awaiting orders.

I dumped my kitbag in the post office and walked back to Kingsbridge in teeming rain. Sure enough, thirty men were clustered on the platform as though for mutual protection, patiently waiting for someone to tell them what to do next. There was a cook, some transport drivers (no transport), some general-duties men, and a full complement of radar equipment operators, the entire crew of a mobile radar station, with no station to go to.

I instructed them to remain where they were while I figured out what to do. At least they were in the dry shelter of the station. I then consulted the station-master, a be-ribboned veteran from the First World War, and explained the situation. The question was where to find accommodation for thirty tired and hungry men on a cold and miserable winter's afternoon. I also told him about the reception in Malborough.

"Look here, lad," he said sympathetically, "I live in Malborough, and they don't trust airmen there, they don't. Why, they'll have locked up their daughters, goats, and everything by now! Best thing you can do is to march your lads to Hope Cove, where there are some hotels, and get them under cover there for the night."

This seemed sensible advice, so I marched the small column to Malborough and there tried to hire a dilapidated three-ton truck to transport their kitbags. However, as the owner of the truck wanted cash on the spot, there was no alternative but to march on to Hope Cove. With capes draped over shoulders and equipment to keep off the worst of the teeming rain, we headed off into the late afternoon. The sunken lane—I would not dignify it by calling it a road—snaked between high banks, and visibility was reduced to an arm's length. Soon total darkness overtook us. After what seemed like hours of slogging through glutinous red Devon mud—for I ordered frequent halts to rest—we slid and slithered down a steep incline onto the grounds of what I took to be a sizeable house.

The house turned out to be an inn with a large yard and a car-park. Leaving the men standing at ease in the rain, I went inside. Beyond the tiny entrance hall was a dark wood-panelled lounge with a blazing log fire to welcome the weary traveller. A warm and cozy atmosphere pervaded the inn, which had the aroma of furniture polish, coffee, and good food. A number of guests were snuggled comfortably in the deep armchairs, some reading newspapers; one well-dressed patron dropped the top half of his newspaper to stare at me with disapproval. On the far side of the lounge a young blonde woman peered suspiciously out of the tiny cubbyhole of the reception desk.

Removing my dripping cape and leaving it in the hall, I squelched in muddy red boots across the carpeted lounge towards her, but before I even opened my mouth she gave me a stern and emphatic "No!"

"What d'you mean, no?" I said furiously at full volume. "I haven't even asked you a question yet."

She lowered her voice, but gave me a defiant stare. "We have no accommodation," she announced.

Startled by the commotion, the hotel guests gave us their undivided attention, but I didn't care. She had my dander up with her superior airs, and I said the first thing that came into my head. "D'you think I'm Oliver Twist asking for more soup? Don't you know there's a war on? Where's the manager? Get me the manager!"

"Here!" said a voice behind me. I turned.

"Bob Halliday," he said. "How do you do? Can I help you?"

He was a well-dressed, middle-aged man with a diplomatic and pleasant manner. When I told him of our situation he tried to convince me that there was lots of tourist accommodation in Hope Cove if I'd care to go a little farther down the road. "Come!" he said, "I'll show you the way," and he walked me across the lounge to the vestibule, where I retrieved my cape-cum-groundsheet from its pool of muddy water in the umbrella stand.

Opening the massive door of the inn, he caught sight of the miserable and bedraggled squad of men waiting in the pouring rain. For a moment he was silent. Then he said, "You'd better bring them inside."

I did, and the men, soaked to the skin, crowded through the entrance hall and followed the innkeeper through the lounge to the warm and spacious kitchen. There they spread themselves out on the warm floor and began removing their sodden clothes. Food was produced, mostly leftovers from

110

the evening meal, and we ate ravenously like starving vultures. Afterwards, I organized a fatigue party to clean up the mess we'd made with our muddy boots and dripping equipment. I was vastly relieved. At least we had a roof over our heads for the night.

I had the first meal of many in a tiny cabin next to the bar where innkeeper Halliday pumped me about what we were doing in the area. I couldn't tell him about radar of course—he would not have understood in any case—but I owed him a debt of gratitude and told him, therefore, of the possibility of starting a fighter base in the area.

"And they've sent you to do that?" he asked in astonishment.

"They're getting a bit short of officers," I lied.

"It seems somewhat odd to me," he said, scratching his head. But he seemed satisfied enough with the explanation. His main concern was the loss of his regular customers with us in the area.

"I wouldn't worry about that," I told him, grinning. "We'll more than make up for the loss of regular customers from what I know of servicemen."

As it turned out, we did more than that, and as for my lie about starting a fighter station, it was nearer the mark than I supposed. Nor would I have believed on that first night that the elegant Cottage Hotel would be our home for some years to come. It also was to become the headquarters for our night-fighter control station, which would cost the Luftwaffe dearly, in both men and aircraft.

I was up early the next morning to find that the rain had stopped. There was a magnificent view of the sea from the hotel: the golden sands, the cove in which Hope Cove nestled, the fresh green countryside. A bright sun in the pale-blue cloudless sky shimmered above the placid water of the cove. The green hills overlooking the headlands were beautiful and tranquil. The hotel sat snugly in the cliff face, and below, that early morning, I could see fishermen at work, pulling in baskets of lobsters. The sharp contrast between

Hope Cove and the stark beauty of Rosehearty was striking, and I knew without question in which place I would rather be.

After a hearty breakfast, I telephoned 10 Group HQ to get direction on how to establish the new station. I drew a blank. Next I called the South Devon HQ of 60 Group and explained that thirty ravenous men would soon eat the Cottage Hotel landlord, Bob Halliday, out of home and business unless we got help.

60 Group responded immediately by arranging for an army signals unit to lay a telephone line to where I indicated, on the property of Soar Mill Farm—a farm which, to this day to my knowledge, does not, and never did, exist. I chose a spot in the middle of a field which I thought suitable for the location of the radar station. On this spot, planted on the grass, an army field telephone was put in a lockable grey wooden box.

Because the headland was a hollowed-out depression, it was not the ideal site for a radar station, which required a flat terrain for accurate height reading. At the same time, it was evident from my survey that no more suitable site was available on Bolt Head. While the men were kept busy killing time by improving the access and taking care to avoid cow-cakes, I kept vigil by the telephone, armed with a box of sandwiches and a flask of tea. I was surrounded by a herd of dairy cows busily exercising their grazing rights, and employed myself by studying the ordnance map to compare it with the topography of the area.

During the afternoon, my vigil was rewarded by a telephone call to advise me that the wheels were turning. The next day I could expect the promised convoy, after which it would be up to me and the crew to make the station operational. Remembering the experience getting the Avebury convoy to the Bristol area, I thought this all sounded too good to be true; however, to be prepared, I kept a man on watch in Malborough. A guide would be needed, for there was no other route for the convoy to reach the headland.

112

He would direct it to the site. I also brought joy to Bob Halliday that evening by presenting him with a pile of ration books sent by dispatch-rider from 10 Group HQ.

Two days went by and no convoy appeared. 60 Group apologized for the delay, but they didn't know what I knew about moving cumbersome equipment. On the second night there was a heavy air raid on Plymouth, a few miles to our west. We could hear the racket of anti-aircraft guns and the crump-crump of exploding bombs, but only as an off-stage disturbance to the tranquil pace of Hope Cove. It was no wonder that the guests in retirement at the Cottage Hotel, and the locals, wanted their haven protected from the horrors of war and such riff-raff as we.

The next day, all was forgiven when the lookout telephoned to announce that the convoy was in sight and that he was climbing on board the leading truck to bring the column to the cow pasture.

We heard the noise of growling engines long before the column hove into view, so we were all ready and waiting with grinning faces to welcome the convoy's arrival. Following hearty slaps and claps all round, we sorted out and lined up the vehicles for operations in such a way that a hillock to the rear would cause the least interference to our view of aircraft approaching from France.

The convoy comprised an operations-room vehicle (a large box-like structure on a Crossley truck), a two-ton transmitter, a 15-Kva diesel-electric generator for power supply, and an aerial trailer. There was also a mobile workshop containing all the equipment we needed to assemble the station and the aerial system. How the convoy negotiated the lane from Malborough to Bolt Head must be left to the imagination, but scuff marks of passage along the sides of the trailers were clearly evident.

We worked furiously and without pause to make the station operational, so that by early evening we had a grandstand radar view of another heavy raid on Plymouth. We were without a radio-telephone connection with sector

113

operations at 10 Group in Exeter, so we could only view as impotent spectators the comings and goings of enemy bombers.

The raid spurred 10 Group into providing us the next day with an Air Fighter Sector (AFS) link now that we were operational. Our Plan Position Indicator (PPI) screen was a 12-inch-diameter picture tube that gave us indication of all aircraft within a radius of sixty miles. To make it fully serviceable, the screen needed to be marked up with a map overlay, so I spent hours doing the artwork with Indian ink, plotting the grid references and features that would make sense of the plots we would pass to Sector Control.

In keeping with the purpose of the GCI system, we would use the plots to navigate the night fighters, despite the fact that we would have to use the VHF sets at Sector Control to communicate with them until we had our own equipment for direct communication with the fighter pilots.

For the next few days, together with GCI Exminster, which was nearer to us than Exeter Sector Control, we practised by following the procedure for night-fighter interception and control during daylight hours. This was no problem, because there was little hostile air activity during the day-time. So, in the space of a few days, our field on the site of the mythical Soar Mill Farm, so recently devoted to pastoral activity, was now a smart little radar station, complete with orderly room, guard room, and other operational features of an RAF station.

We had no inquisitive visitors from across the water, but the lack of decent camouflage concerned me. So, with the help of an army unit I had discovered stationed at Kingsbridge, we draped sacking over the vehicles to disguise the appearance of the station and to make it merge with the landscape when viewed in silhouette at a sharp angle.

By early March we were ready and eager to do business. All we needed were customers. Bolt Head was no longer the figment of some planning officer's infertile imagination. The station was real and I was proud of the achievement, because, without the on-site involvement of planners, surveyors, and a host of staff officers, GCI Bolt Head was a reality—and we had made it happen. If credit was due, it was due to the efforts of thirty men, not one of whom was higher in rank than sergeant.

10

NIGHT ATTACKS

Hope Cove on the headland called Bolt Head is an out-of-the-way spot. Overlooking the Salcombe Estuary, the cove was the kind of place where buses ended their journeys, where summer visitors came for a paddle, a picnic, and then went back to where they had come from. For the rest of the year, Hope Cove was a quiet little community that existed without close contact with the outside world.

The Cottage Hotel, sitting snugly on a hillside within walking distance of the village centre, overlooked the sea. It was a neat-looking building with white walls and black trim, a low porch entrance, and small-paned windows. Against the bright green hillside, the hotel was starkly visible in clear weather for miles out to sea. Anyone who has ever visited this delightful place will agree that the Cottage Hotel is perfect for a quiet, relaxing holiday.

In 1941, untouched by the war until we arrived, the hotel catered to a few resident guests and local clientele, who congregated at the lounge bar for the evening. The lounge was well furnished with deep, relaxing leather armchairs, writing-tables, a large stone fireplace, and rich deep-pile carpets.

The bar had been built from the remains of the four-masted schooner *Herzogin Cecile*, which ran aground off Hope Cove shortly before the war. The entire cabin complete with benches, tables, and portholes was removed and fitted into the decor of the hotel. The story was that the

Jack Nissen in 1941, after a training session in Scotland, in the full-dress regalia of the Gordon Highlanders.

A mobile Würzburg short-range radar for anti-aircraft gun and searchlight control, used by the Germans early in the war.

Photograph taken by the Germans with a telephoto lens from France. It shows the four small aerials of two CHL radars as well as the 240′ receiver mast of CH Dover.

(Left) Sir Robert Watson-Watt, the inventor of British RDF, which was subsequently called radar.

(Right) Air Vice-Marshal Sir Victor Hubert Tait, KBE, OBE, CB, RAF Director General of Radar and Signals. It was to him that I reported after the Dieppe Raid the fact that the German Freya was now a precision radar.

Captured Canadian soldiers at about 1:00 p.m., August 19, 1942, at Pourville. The South Saskatchewan Regiment came ashore on the beach just behind the house at top left centre.

Intermediate/Mobile GCI at RAF Station Bolt Head. This night-fighter control station could navigate and direct many fighters at the same time.

Jack Nissen and Willi Weber, the radar officer in charge of Freya 28 during the Dieppe Raid. After the war, Weber became a General in NATO.

Field Security Sergeant Roy Hawkins who accompanied me on my cross-country travels and who swam out with me at the end of the raid.

An airplane with the magnetron-equipped H2S/ASV III radar. The switch box to the right is a primitive analogue computer. This radar enabled us to win the Battle of the Atlantic in 1943.

The magnetron, our "secret weapon", was the heart of the most powerful radar transmitters developed up to that time.

Mid-channel memorial service (1985) for the men of the 1st Destroyer Flotilla who died during the Dieppe operation.

cabin was a place of debauchery when it was afloat, and this was easy to believe when one's equilibrium was disturbed by over-drinking. In our charge, though vast quantities of cider and beer were consumed, the bar, the hotel, its fittings and decor were at all times respected and maintained in excellent condition.

There were a number of guest bedrooms on the second floor, and the best of these was the "Blue Room", which I occupied during the whole time I ran the Bolt Head GCI station. The room had a blue decor: chintz curtains, blue blankets, and pale-blue eiderdown; Georgian polished walnut furniture and thick-pile blue carpeting; even a blue-tiled bathroom with an ultra-modern blue sunken bath—such blissful luxury.

To enjoy the mild and pleasant Devon weather under sunny skies was a wonderful change from the raw experience of plunging through the snow of Aberdeenshire to do a tour of duty. Now I could walk along the sea front when there was time to spare. Apart from having to keep clear of the mined sections, the war, on these expeditions clambering lichen-encrusted rocks along the shore, seemed far away.

Shortly after we arrived, the regular guests moved out and the owner-manager, Bob Halliday, took up residence along the coast in the Thurleston Hotel, of which he was a part-owner. The Cottage Hotel, for the duration of the war, was then taken over and was run by the RAF. This arrangement was made on the strict understanding that the government would pay for any damage done during occupancy by service personnel.

One hardly questioned the protocol developed by which I, now a flight sergeant, organized the base. There were administrative snags to be overcome, of course, and for these one had the problems dealt with at arm's length by 10 Group HQ. Officers at the administrative level rarely put in an appearance; they seemed content to leave the day-to-day operation of the station to me. The fact is, it was to be operated with a minimum of personnel, so that, if successful, the unit would be the model for others that could be shipped overseas en bloc when needed. This was the broad plan. How it would work out had yet to be determined. The transition of our virtual takeover of the Cottage Hotel went smoothly and, from the morning after our arrival, was never in doubt.

Having set up the GCI equipment, we got acquainted with 307 Polish Night Fighter Squadron, which was to work with us to develop our ground-control procedure. The first priority was to train GCI equipment operators and to improve our performance of directing fighters. Two Spitfires of the 317 Polish Squadron, sometimes more, were assigned. There already were two Polish squadrons in the area: 307 Beaufighters Squadron, operating out of Exeter, and 317, a Spitfire squadron. Within a couple of weeks the aircraft of 317 Squadron were using the grass strips alongside the installation as temporary landing facilities. It seemed pointless not to have the aircraft and radar control together, so, during May, we arranged for a work gang to lay the mesh for two runways, strategically set out to take advantage of the prevailing wind. The idea was to develop a prototype defence unit consisting of radar control, aircraft, and servicing personnel.

The Poles liked our situation and location, probably as much for the hotel as for anything else. Being survivors of the 1939 invasion of their country, they were a brave and fearless bunch. They made their way to Britain in as many ways as there were men in the squadron.

There was, of course, the problem of language between us, since they were newly arrived, their squadron was newly formed, and most of them barely understood English. Some had escaped from Poland through Rumania overland, some by diverse routes through the Middle East. Many of them succeeded in escaping from Polish airfields in their aircraft, and flying them to France, where they were taken into the French Air Force. Following the collapse of France, they did another flit like unwanted tenants, flying their machines to Britain, where they were received by the RAF with open arms.

This could be risky. Under the control of fighter sections, the Polish fighter pilots took their directions from sector controllers who, naturally, spoke English. Instructions were frequently misunderstood by the Poles, which often led to serious situations. In combat with the enemy they lapsed into Polish, which left their ground controllers utterly in the dark. To some extent these difficulties were overcome when the RAF, in June 1940, quite sensibly trained Polish ground controllers for Polish squadron operations. So, in exchange for minor administrative difficulties, the RAF got the benefit of a large number of experienced and fully trained fighter pilots at a time of the greatest need.

Even before the Battle of Britain waned, Dowding began looking around for high-calibre pilots to fly the new Defiant and Beaufighter aircraft fitted with the first AI units for night work. Pilots with experience in instrument flying, who were brave enough to undertake dangerous night assignments when there were few or no aids to flying, were needed. When the suggestion was put to the Poles, there was no shortage of volunteers, and 307 Polish Night Fighter Squadron was formed.

The really odd thing I noticed about the Poles when I first met them at Exeter was their age. Their squadron leader gave his age as twenty-nine when I inquired, yet I'm certain that he was more than ten years older. His radar observer was at least fifty, but tactfully reverted to "No spik!" when the topic of age was raised. And although these "old men" were twice the age of the average radar operator, together we made a winning combination in the dicey operation of defending the night sky over our patch of Britain.

Within two weeks of my arrival at Soar Mill, or RAF Station Bolt Head as I named the station, we began working with the Mk.II Spitfires of 317 Squadron. My informal introduction to them occurred one morning at 7:30 when Bill

Powell, my orderly-room sergeant, reported from the guard room, "You've got a deputation here, mate. It's your friends the Polacks. They want to speak to you before you go back to the billet."

Pilots were not allowed in the radar compound, so I hopped on the BSA motorcycle I used for transport and roared around the "drome" perimeter to the Nissen hut that served as an orderly room. The four visitors were uncommonly quiet when I went in, standing properly at ease as though they were on parade.

I said "Good morning" to them in Polish, which they appreciated, and went into my office, where Powell, lounging in my chair, sat with his face wreathed in a big grin. He was older than I, an ex-Shanghai policeman chosen to police the installation on account of the need for strict security.

"What's it all about?"

"You'll find out," he said.

"That's a lot of help."

I had him show them in. The delegation, led by a burly flight sergeant, marched in and lined up in front of the desk. It was comic the way they entered, marching stiffly in step, and coming to a halt to gaze at an imaginary spot above my head. I wanted to smile. The flight clicked his heels respectfully and gave a little Germanic bow. I asked him what I could do for him and, in fractured English, he answered.

They had been at Bolt Head for two months without action, he said, and now they wanted permission to attack the enemy over France. I pointed out that France was a long way away and that the range of their aircraft was limited. Furthermore, four aircraft could make very little difference to the progress of the war. Anxious not to dampen their offensive spirit, which is vital to every pilot, I said they might well get there, but not back. I was thinking about the fuel capacity.

The flight sergeant thereupon produced a carefully prepared flight record to show that two of the squadron aircraft had stayed aloft for well over two and a half hours, and still had enough fuel in their tanks for more flying.

I hadn't realized that their Mk.II Spitfires had that sort of endurance, and I did not want to suggest what was foremost in my mind—namely, that they had "cooked the books". The four aircraft and pilots were only on loan to Bolt Head for controller-training, so it was none of my business. I told them that their squadron commander in Exeter was the one to authorize operations over France.

The sergeant exploded and said his chief didn't have the guts to fight. Knowing that their commanding officer had crossed Rumania and fought his way across Yugoslavia to escape, I ventured to disagree, but they were not to be mollified. Why had they come to me? The leader said that, because I pulled many strings, they were convinced that I was not really a sergeant but of a higher rank. They knew I had more power than anyone else at the aerodome.

The Poles missed nothing. They were well aware that there was a constant stream of senior RAF officers, their hats dripping with gold braid, parading in and out of the operations room. With visitors of such rank, they reasoned, the station was something special, and we therefore had to have something to do with locating and tracking aircraft. More to the point, they knew that the station was my baby, and they deduced from that, erroneously, that I had the power to move the authorities to sanction offensive action. This, of course, was not true. I had no such power.

To mollify them, I said I'd take the matter up with 10 Group HQ. Only with this assurance were they willing to depart. Outside, I heard them break into voluble chatter, as though they thought it a *fait accompli* that they would soon be carrying out raids over the water. There was a grain of truth in their assumption.

Towards the middle of 1941, the RAF mounted a series of attacks code-named Rhubarb. The attacks were made on France from the Sussex and Kent coastal regions, because they were near to the continent. A Rhubarb was a search-and-destroy mission in which aircraft crossing the Channel would be free to shoot up vehicles and installations at will. It was one thing to raid from forward airfields not more than twenty or thirty miles from the target area, but Bolt Head lay three times that distance from the French coast. Since we knew by this time of the Germans' Freya warning system, we also knew that the chance of flights from Bolt Head meeting a reception committee was strong. Therefore, the prospect of fighting over France with a long outward and a long return journey, and possibly with a damaged aircraft, made attacks out of Bolt Head extremely dicey.

But the Poles were a rambunctious lot. Some of them had hundreds of flying hours in their flight logs and were confident they could attack and return to Bolt Head unscathed. This was perhaps true, but a pilot error in assessing wind drift could easily put them in the Atlantic off the Cornish coast. Against my own better judgement, I contacted an associate at sector control and told him of the endurance tests the Poles had carried out, and of the results obtained. In response, he told me that Fighter Command at Stanmore had asked 10 Group to explore the possibility of conducting Rhubarbs from Bolt Head. If this was possible, it would tie down German fighters that would otherwise be transferred to the eastern battle front, where, at that time, Russia seemed to be near defeat. The prospect of mounting a raid was exciting.

A delegation of senior officers arrived to consider the situation, and gave us the go-ahead. The Poles were jubilant. To their delight, it was decided to conduct an operational test. Four aircraft were to fly on a course in the mode I directed, and the average fuel consumption during the operation was to be carefully computed. There was to be no combat with the Germans. This was to be a trial trip to France and back, no more. On their return, their fuel was to be checked to decide if future attacks could be made using the Mk.II Spitfire. The one stipulation that raised a moan was compulsory attendance at the English lessons I'd arranged because of the difficulty some pilots still had communicating with the ground-control operators. To soften the blow, the lessons were given by a pretty WAAF who came from Luxembourg and spoke fluent Polish. She was soon nicknamed the "Countess of Luxembourg" and treated in fun with full pomp and ceremony as though she really were of royal stock.

We had known from as early as November 1940 that the Germans had radar installations in occupied France. Pictures of a unit located on a headland at Auderville, near Cherbourg, directly opposite Bolt Head, from which I received powerful signals, strongly resembled one of our early CHL-type radars. From its long wavelength it was fairly easy to assess its performance. It was similar to my old station at Rosehearty, and my experience there indicated that it could not "see" aircraft below 500 feet at close range with reliability. This meant that halfway across the Channel the Spitfires would have to descend to a low altitude to cross the French coast.

Without divulging much technical information, I explained to our would-be bandits the way in which they should fly. In our area at that time there was no air-sea rescue service, so it would be tough luck if they ditched. It was important for them to maintain sufficient altitude on the return journey to enable us to make and maintain radio-telephone contact. Finally, I emphasized that they were not to provoke a shoot-out with the Luftwaffe; the excursion was to be a proving flight and no more. They nodded in agreement with all I said, grinning like schoolboys anxious to leave for the summer holidays, and if I caught the odd conspiratorial exchange of glances there was little I could do about it.

Following the usual checks, they took off. We followed them on the screen until they were about thirty miles out. Then, faithful to my instructions, they dropped to a low altitude to get under the German Freya radar and disappeared from our radar view. With the loudspeaker volume turned up high, we then sat back and waited, listening to the steady hiss and static of the radio-telephone. The quality of our VHF radio was excellent. With an aircraft at 8000 feet we could speak clearly and distinctly at a range of about 80 miles. However, at low altitudes the range was severely restricted.

We watched the time crawl by on the sector clock and betrayed our anxiety by drumming our fingers on the console and grimacing at one another; we were like mothers whose children were spending their first afternoon at school. Seventy, eighty, ninety minutes went by and we began scanning the radar screen for a sign of our charges, but the first indication that all was well came first from weak noises on the loudspeaker. Our friends seemed happy enough; they were chattering to each other in Polish.

Then their flight leader called in in broken English. Next we picked them up on the screen, flying high and making a beeline for Bolt Head. In no time they were overhead, merrily buzzing the station. My ears pricked up at the sound they were making, a high-pitched whistling noise that comes from a Spitfire when the fabric covers protecting the machine guns are punctured. That was not part of the deal. There was not supposed to be any contact with the Luftwaffe on this mission.

I inspected two of the parked Spitfires on my way to the dispersal hut and confirmed that the guns had been fired. One of the machines had superficial damage to its tail section. Even before I went into the hut I could hear the high-spirited, high-speed exchange of Polish, punctuated by the guttural Polish equivalent of "Rat-a-tat-tat!" The chattering stopped as I entered, and they greeted me with a quartet of sheepish, triumphant grins, the rogues.

A maintenance mechanic said there was still fuel in the tanks. When I asked how much, he mumbled something obscure. The flight commander, with a smile from ear to ear like a slice of watermelon, said they had had a good flight, "Werry good!"

"But there was to be no contact with the Germans."

The excuse, extracted by a controller who was intent on debriefing the flight, was that they encountered heavy cloud shortly before they made landfall, and before they knew where they were they were attacked by an equal number of Messerschmitts and were forced into a fight. No, they didn't know how long the fracas had lasted, but by all accounts it went on for at least a quarter of an hour.

It later transpired that they had already reported to their Polish intelligence officer at sector control to say they had damaged two Messerschmitts and possibly destroyed a third. It was exciting news, which we celebrated with the coffee and vodka the lads produced. I drank the coffee and sighed, relieved. All's well that ends well, I thought. They were home safe and sound, so all was forgiven.

There was one discordant note on the affair, and it was a matter for serious concern. Asked if he had followed my instruction to fly low on both the outward and the return legs, the flight commander said he had decided on the reasonable height of 5000 feet above sea level coming back because they were low on fuel, having used it up in combat. I was dismayed; 1500 feet was the maximum height to keep under the minimum Freya surveillance height over the Channel. The news could only mean one thing: the Germans would now have a track on their plots that pointed like a weathervane to our station.

Hurrying back to the control tender, I promptly contacted Lieutenant Hamilton-Smythe, Officer Commanding the detachment of the Queen's Own Regiment at Kingsbridge, and asked him to speed up the improved camouflage he'd promised for the vehicles. On the headland, the installation still stood out too conspicuously for my peace of mind. If the enemy did pay us a visit, the safety of the night-fighter control equipment would be jeopardized, and that equipment and its use was the basic justification for our existence. Hamilton-Smythe accommodated us by promptly sending a detachment of men from his unit to drape our trailers in camouflage nets.

We needed to improve the operating range of the Spits. I had heard about a new fuel tank that could be discarded when empty. Indeed, Ventral tanks, as they were named, had to be jettisoned before the aircraft was used in combat. I made inquiries at HQ and was told that long-range tanks were available and would be fitted; they were, by a special crew sent from Exeter a few days later.

Group HQ was so pleased with the results of the first Rhubarb from Bolt Head that a decision was made to move the entire 317 Squadron to the station. Very quickly, construction crews arrived to expand the aerodrome. More Nissen huts were thrown up, the perimeter was enlarged, and high-blast walls covered with turf were built for the protection of parked aircraft in the event of attack. The facilities of the station were increased to feed, house, and maintain the hundred or so men and women needed to support a full fighter squadron.

The Poles were capable and enthusiastic, and were prepared to put up with conditions that RAF British staff would perhaps not have stood for. The hard-working crews that came with the squadron were also caught up in the enthusiasm exuded by the pilots, and cheerfully accepted the less than ideal facilities that we offered. Bolt Head after all was no holiday resort, despite its close proximity to Hope Cove.

It took us less than two weeks to prepare for the first squadron Rhubarb. After a short briefing in the early morning, the entire squadron got on its way and soon vanished from our radar screens. Once again we watched the clock, hoping and praying our Polish gun-slingers wouldn't be jumped by the Jerries.

I had warned them repeatedly not to use their radio transmitters until they were in combat, so was pleased to note that they observed this instruction. Any radio contact before battle would help to alert the enemy radar warning system.

One couldn't help admiring the spunk of these Polish cavaliers. We had developed a close relationship with them over the past few weeks, and now, out on their first squadron Rhubarb, we felt deeply for their bravery and courage. They were our champions and we their squires, experiencing the thrill of mortal combat. I also have to admit that I had developed a healthy respect for the enemy. The disgusting Nazi doctrine and philosophy aside, they were no fools and could not be over-estimated.

Odd snatches of guttural Polish pierced the static of the radio loudspeaker. By our reckoning, the squadron was well over the French coast, which meant that they were probably in contact with the Luftwaffe at a fairly high altitude. If not, we could not have picked them up as clearly as we did. The Mk.II Spitfire did not perform well at high altitude, but we knew that the Poles were spoiling for a fight, even if it had to be at a disadvantage.

Then came a pilot's distress call in broken English, to say that he was in trouble and was making for home. Almost immediately two more admitted they were also in trouble. We acknowledged receipt of the messages and wondered if they would make it back to Bolt Head. They had a long trip ahead of them. The thought of pilots in trouble was a frightening aspect of the whole business, but we could only sit and wait. They had to conserve fuel. We knew that our pilots had developed a technique of gliding from high altitude using low engine revs. It had proved economical and successful for the original four pilots, and we hoped the others knew enough to follow their example.

There was another long wait, interrupted by serious talk on the radio intercom. Even though we didn't understand Polish, we could tell by the tone of voice that there was a good deal of concern. Then a faint indication of a broad IFF pulse came on the screen, indicating an aircraft in distress. IFF units could be used to broadcast a distress signal if the pilot ran into trouble. The WAAF operators peered anxiously at the screen. With thirty miles to go, the first of our friends was on the way in.

I took over a head-set and called. The pilot's engine was overheating and showing a RED indication. He would, he said, have to bale out. Another, leaking glycol coolant, was confident he could make it back to Bolt Head.

It was all very exciting and dramatic, and the spirit infected everyone. The pilot with the glycol leak was as happy as anyone could conceivably be who was flying a damaged aircraft miles out at sea with precious little fuel remaining. He called again, saying he had sighted land and could he come straight in.

"Roger! It's all yours," I said, jubilant with hope he would make it.

Large rocks on the cliff edge, to the right and the left of the runway, served as visual markers for the start of our airstrip. Coming in low, he lifted his Spit clear of the cliff edge and made a direct approach. Not till he was committed to landing did we see that his undercarriage was still retracted. He was already feeling for the ground when his propeller touched the deck and the aircraft slithered, scraped, and skidded in a straight line along the strip before coming to rest. Its propeller was hopelessly twisted and the underside was damaged, but the pilot, unhurt, and looking sheepish, emerged from the cockpit with the help of the fire-tender crew that had sped to his aid.

If it had been possible for him to carry through the customary "port orbit" (a left-hand circuit of the landing strip) before making his approach, we would have seen the undercarriage retracted and warned him. Now, though our first chick was home, we had a bent Spitfire on our hands.

The bulk of the squadron reported in, flying as a squadron with their squadron commander leading them in triumph. News of the other two aircraft in distress was not as good. One pilot, after a final desperate call, had taken to his parachute almost ten miles out to sea. A Lysander search aircraft was dispatched to look for him, and when it located him, a boat from Salcombe picked him up. The third pilot, now very low over the water, was anxiously trying to make a landfall.

This was dangerous. He could be too low to parachute and have too little power to climb to cliff height. It was always the older pilots who would not, or did not want to, get their feet wet. In this case we could do nothing but wait and pray.

A short time later there was a call from the coastguard a mile or two away to tell us that one of our aircraft had gone down. He had crashed into the cliff face and been killed. It was a sad ending to an exciting and almost faultless Rhubarb, our first in squadron strength. On the basis of the debriefing intelligence, of the surprise achieved, and of enemy aircraft shot down, the mission was highly successful.

11

NAVIGATIONAL RADAR—THE TIDE TURNS

The Rhubarb operations out of Bolt Head grew in magnitude and intensity. The constant drama in which we lived unhappily brought a steady stream of funerals, although more often than not there was little to bury. The loss of precious lives increased following the appointment of Air Vice-Marshal Leigh Mallory as Chief of Fighter Command, for he was more aggressive (or reckless, depending on one's point of view) than his predecessor. Heroic "orders of the day" were issued, praising the "victories" of our fighters over France. He pushed and prodded Fighter Command into taking more and more offensive action, which in many ways was to his credit.

We had numerous pieces of new equipment, but some of them caused problems. For instance, the new wing tanks added to Spitfires to increase their operating range also increased the crash rate on take-off. The extra Ventral fuel tanks were supposed to provide fuel for the first part of the flight, and then be jettisoned before combat regardless of the fuel in them. The procedure was for the pilot to take off using the normal petrol tank and to switch as quickly as possible to the Ventral tank in order to use that fuel first. Many pilots took the instruction too literally and switched tanks before leaving the strip. As a result, there was a series of fatal end-of-the-runway crashes and burn-ups. It was some time before we determined the cause.

What happened was that, on switch-over, the engine would falter because of an airlock in the fuel lines before it picked up again. The loss of power in a light aircraft that was fully loaded but not yet airborne led to disaster. The pilots stood little or no chance of surviving full-tank burn-ups. When the cause of the trouble was recognized, we corrected it by instructing that change-over be made only after the aircraft had reached a considerable height.

Bolt Head had begun as a night-fighter control station, which meant that I could sleep during the day. With the development of Rhubarbs and the arrival of the entire 317 Squadron, there were night and day operations to supervise. With the onset of good flying weather, from June 1941 on, our days were crammed with excitement, so I counted myself lucky to snatch three hours' sleep in every twenty-four. It is amazing what young minds and bodies can stand. Despite my strange way of life I remained fit.

There were still night-interception operations to attend to and faults to correct in the AI-equipped Beaufighters. Partly as a reward, I suppose, for success in reducing an annoying "jitter" on the Beaufighter picture tube, I enjoyed my first stint of operational flying as a radar observer from Exeter. However, there was a little more to the jaunt than a simple reward for good work. I have explained the need for more detailed interception information once the GCI controller had put an aircraft on a bandit's track. In the attack aircraft, erratic and bounding pictures appearing on the radar screen made it difficult for the aircraft radar operator to direct the pilot to the target, and this was the problem I was to work on when on my jaunt.

The Mark IV AI installed in our Beaufighters had two cath-ode-ray tubes with which to pinpoint the target and to position the night fighter. One tube indicated the horizontal deviation from centre, whether the target was right or left of the aircraft; the other gave the vertical indication. By watching the two pictures, the airborne radar operator could guide his pilot by means of a running commentary. The radar signals received on one tube from the aerials on the aircraft's wing-tips were for the horizontal signals, and on the other, from beneath the wings for the vertical indication, were manipulated by an electric motor that rotated a series of aerial switching contacts. In this way, using the two pairs of signals, the radar operator obtained a continuous picture. Adjusting and servicing these tiny phosphor bronze switch fingers was an art I had learned while working at Rosehearty, where the same principle was used for obtaining a bearing on ships at sea.

The period of practice interceptions in the air was almost up, and so far I had not so much as smelled a target. The week before had been one of intense activity for Bolt Head, as single intruders tried their luck in disrupting our operations. In comparison, "business" during my week aloft had been bad, mostly owing to atrocious weather conditions over France. The weather on my last night at Exeter before returning to Hope Cove was awful, with heavy overcast and slashing rain.

Despite this, I was determined not to miss the last opportunity for night flying, even though the prospect of a contact was poor. As we climbed above the cloud cover, the starry sky appeared through the astrohatch and became bright enough for one to see. The Milky Way, so thick and rich with stars, served as a backdrop for the visual identification of targets. Up there, in the night sky, the stars seemed closer somehow, and friendly, like beacons to light our way.

I had the brightness of the AI set turned low to retain some of my night vision, and, under the control of Exminster GCI, we began our practice intercepts. I muffed the first one, but quickly restored my self-confidence with the next. Once I had the feel of the equipment, and the confidence of the pilot, one of my Polish friends, we intercepted our friendly bandit easily and followed him through a series of complicated kinks. Even with the improved split-switch operation, the AI trace showed its characteristic jitter, yet it was now usable at the shortest limit of the unit's range.

The minimum range to which a pilot could be directed on to the target was most important. In poor visibility it was touch and go to be able to see another aircraft in the night sky, even as close as a hundred feet. With the AI working well, it was now possible to get the pilot within the necessary 500 feet, thanks to Alan Blumlein's modulator.

As the night wore on, we "played" over the Channel, taking turns with our companion in being "fighter" and "target". The rules of the game allowed the target pilot to choose his height between the limits of 10,000 and 15,000 feet, for the GCI station on the ground also needed to practise height reading. We were in the middle of the English Channel when the exercise ended, and we were about to make tracks for home when suddenly the GCI controller below told us we had a "customer". He directed us to an incoming aircraft, flying high above us and travelling at a comparatively slow speed.

An agreement with our own Bomber Command required that returning aircraft stay below 8000 feet. Any aircraft flying above the "angels eight" altitude was, as a consequence of the agreement, to be considered hostile. To avoid attacking our own bombers, though, it was necessary to positively identify the intruder before bringing the fighter's cannons and machine-guns to bear.

A series of vectors from Bolt Head GCI brought us up behind the invader just off the coast, somewhere near Torquay. We finally made our own AI contact at the maximum range and closed in. He was high to starboard. The controller had given us his correct speed and bearing, so we approached his rear end warily. As the range decreased, my adrenalin ran and my excitement mounted. This was it; my patience of the past week was to be rewarded.

"To our right and half a mile ahead," I told the pilot.

I couldn't leave the tubes to look outside in case the raider spotted us and took evasive action. By watching the tubes, I could ensure that we stayed on his tail. Closer and closer we drew, until at last the target disappeared into our own transmitter pulse.

"I cannot see him," said the pilot.

With his permission, I hoisted myself to the astrohatch and peeped out of its small aperture. Just as I looked out, a stream of coloured floating "neon" lights zipped by to be lost in space. The neon lights I saw were tracer bullets being fired at us by the target aircraft's rear gunner. Then the pilot saw the quarry and closed in to identify it. Quite suddenly we saw the chunky outline of a Wellington bomber—and once again the rear gunner opened fire on us.

"Vrump!" We dropped away to the left to get clear.

I picked up the Exeter radio beacon after that and we headed for home. No, we didn't land without incident. We had just made an otherwise perfect landing in a gusty cross-wind and slashing rain when I was suddenly flung against the bulkhead and held there by the movement of the Beaufighter.

"Not to worry," said the nonchalant pilot when I'd recovered and we were taxiing towards the apron. "We only skidded off the runway."

He might have been all right, strapped in place as he was, but the slewing gave me a nasty turn. I hadn't had such a jolt for a long time and decided then and there that flying capers were not for me after all. I preferred to return to my less hazardous duties at Bolt Head.

This was but one of the many experiences I had while stationed at Bolt Head helping perfect the GCI system. The improved results in terms of aircraft shot down at night speak for themselves. During the last six months of 1940, night fighters claimed an average of 4 bombers a month. This was due largely to the exceptional abilities of pilots such as "Sailor" Malan, the South African, who held the top score of 4 bombers for the six-month period. But from the time we had GCI Bolt Head in operation, results improved dramatically. By March the score was 10. By the end of May, when King George visited Exminster, the combined score from January 1941 to May 1941 totalled 96.

The phenomenal success was as much due to tightened-up procedures as to improvements in the equipment. These procedures included speedier transmission of information on intruders from Stanmore (using the CH-CHL network) to sector stations, and from the sectors to the assigned GCI station. The system functioned like a well-oiled machine, with periods of heightened drama for the GCI station, as can be imagined.

In late May 1941, a visit by the King to 10 Group was arranged, so that he could witness the interception system in operation. The Avebury GCI, relocated to Exminster, was chosen for the demonstration, and, following the usual parade and lunch, on this occasion in the 604 Squadron mess at Middle Wallop, the royal visitor and his retinue headed for the River Ex. The station, located thirty miles north-east of Bolt Head, was constructed in a swamp, which gave the station's radar system certain advantages over other stations: the flat reflecting surface of the swamp gave Exminster almost perfect height calibration.

Flight Lieutenant John Brown was controlling an interception exercise when the King and his equerry, Wing Commander Peter Townsend, squeezed into the tiny control centre. To Brown's right there was a small telephone exchange. To his front stood the two radar consoles with their large screens. The PPI was for use by the controller to watch the fighter and any target, and this took the form of a superimposed map of the coast sixty miles around, south of Exeter. The other screen served to give the heights of aircraft, fighter, and target if a customer was available.

The station controller used the Plan Position Indicator and a telephone head-set to speak to the pilot. On the controller's right sat the plotter and the height reader. The plotter's job was to provide the aircraft navigator, through the controller, with the plots on the fighter and its target.

Plots were also passed to the Sector Operations Room, where they were registered on a large mapping-table for their information.

Once the GCI controller had taken over the night fighter, the GCI crew was entirely responsible for the Beaufighter's navigation and safety. It was a hazardous responsibility, because, if the fighter got too close to the bomber, or overtook it as sometimes happened, the fighter was in danger of being shot down. In the early days of GCI this had happened.

On the occasion of the King's visit, John Brown explained the procedure. The King, being a pilot himself, understood the navigational processes to be followed in achieving interception. He sat on a stool behind the plotter and watched the team conduct a practice interception in the English Channel fifty miles away. He was fascinated.

During these exercises, Bolt Head and Exminster took turns controlling the fighter and the target. To disguise instructions from enemy eavesdroppers (though it didn't fool the Germans for a minute, I later discovered), we developed our own jargon: "customer", to which I've already referred, was an intruder; GCI was the "shop"; and "business" was the actual work of intercepting the enemy, as opposed to practice intercepts. Townsend had already explained this to His Majesty.

Tea was taken in the tiny cabin as the evening wore on (the King was a regular sort of bloke to take tea with the RAF and WAAF operators; it was all very chummy), but it looked as though he would have to be satisfied with the practice intercepts. Then, at ten o'clock, the atmosphere became electric when Stanmore reported an unidentified aircraft, an "X" raid, approaching the South Devon coast. It was not showing IFF. CH West Prawle, near Salcombe, had plotted the potential customer, which they read as flying at 15,000 feet.

The first pinpoints indicated that the intruder was under 80 miles south of Exminster, so the GCI scanned that sector and waited for the target to appear on the edge of its screen at the 60-mile limit. As the unidentified aircraft appeared to be flying the Knickebein beam from Cherbourg, we undoubtedly had a customer.

Bill Pratley, operating the height tube, which had a 90-mile range, saw the blip first. There was tense excitement when the King, Brown, and the plotter saw the blip appear on the very edge of the PPI and all spoke at the same time: "There he is!" "I see him!" "There!"

The pin-point was read off and the crew navigator started his stop-watch. This was a momentous occasion for the whole of the radar network, because, being in radio contact, all stations in the system got the news and quickly became aware of the drama now playing itself out in the middle of a South Devon swamp in the presence of the King.

Calmly—there was ample time—the controller asked sector control to call the first night pilot waiting to be directed on to a raid. Peter Townsend had once commanded 604 Squadron at Middle Wallop, and, by the luck of the draw, one of Townsend's own pilots, Flight Lieutenant John Cunningham, was assigned. Brown was pleased to have a star performer for the intercept, because he knew Cunningham would pull it off if humanly possible. Cunningham, already airborne and over the Channel, was told to change his radio channel and call "Starlight", Exminster's call sign.

Immediately the two were in friendly contact, as though enjoying a casual telephone conversation.

"I have a customer for you. Starboard Vector 270."

"Roger!"

If controller Brown could contrive to manoeuvre the Beaufighter to get the Bold Head land mass in the background, the fighter would be invisible to the bomber. If, however, he brought the Beaufighter behind the bomber when over the sea, the customer's rear gunner might pick up the fighter's outline against the moon-reflecting water. John decided to wait until both aircraft made a landfall. The fighter was at 12,000 feet; the bomber at 15,000.

High above Bolt Head, Cunningham was directed to creep

in below and behind the unsuspecting intruder. The fighter's AI operator located the customer and took over close-quarter control. Then the hawk-eyed Cunningham made visual contact, seeing the customer glistening in the moonlight. It was a Heinkel III, carrying a crew of five, sublimely winging its way toward the night's target.

Cunningham crept to within 400 feet and steadied his aircraft—a flying gun platform with the fire-power of a small battleship. Warning Peter Rawnsley, his radio operator, he lifted the nose and went in to attack.

The closing contact was clearly visible to the watchers below, both to us at Bolt Head and to the cubby-holed group at GCI Exminster. We gazed, fascinated, as the two aircraft merged into one directly overhead. Watching the blips close on the screen produced its own nervous tension. Brown asked the King and Townsend if they would like to step outside and witness the *coup de grâce*. They did, stepping carefully down the trailer steps, through the canvas light-trap, and onto the narrow walkway between the transmitter vehicle and the control centre.

The steady hum of the Heinkel could be heard distinctly against the powerful roar of the Beaufighter. It seemed impossible that the German air crew did not know that oblivion was less than a minute away. Suddenly there was a staccato roar of cannons, a flash of light high in the sky, and, moments later, the spine-chilling whine of an aircraft in its death dive.

Now that it had happened, the King, according to John Brown, was horrified at the cold-blooded destruction that had taken place before his eyes. A kind and gentle man, he was visibly shaken. Subdued, he returned to the operations room to thank Brown and his crew for the explicit demonstration of science at war. John said that he saw the sadness in the King's eyes when he remarked that he looked forward to the day when science could once again be used to help the human race and not to destroy it.

On September 22, 1942, Air Marshal A. T. Harris, who was to become known as "Bomber" Harris, was appointed as Chief of Bomber Command. Previous to the appointment, Harris was the Deputy Chief of Air Staff. He had little confidence in radio navigation aids and, as early as February 1941, had dismissed them in favour of "dead reckoning" astronavigation, which he considered perfectly satisfactory for conducting bombing raids over German-occupied territory.

He was not an easy man to deal with. His manner was abrupt and often rude with those with whom he disagreed. Even when he took over Bomber Command, with the express purpose of introducing the new policy of bombing by radar aids as instructed by Churchill, he was no easier to get on with. When Watson-Watt proposed the use of a new system, Trinity, to help navigators more accurately locate the target, Harris crudely told him what he could do with his boxes of wires. Harris did what he was told to do all the same, despite his intractable manner. Bomber Command had a problem, and it was the task of Harris, as Chief of Bomber Command, to help solve it.

Towards the end of 1941 a series of air reconnaissance photographs, which followed some disturbing intelligence reports, had proved that our bombers were hopelessly off target. Further, losses in aircraft and air crew had been very high throughout 1941, and the reliance on dead reckoning and astronavigation had made the operations of Bomber Command a costly and bloody farce. It was this heavy loss in men and machines, not to mention the failure to hit the targets, that brought about the decision to seek a better navigation system.

Trinity, the navigation system Watson-Watt proposed to Harris, had been discussed and argued over since the Bawdsey days. Robert Dippy, the inventor, had first proposed the idea in 1937. Now, with aircraft navigation becoming a top-priority item, Dippy was put in charge of the project, and

Trinity (later known as Gee) was first introduced in March 1942.

The principle of Dippy's Trinity system of navigation was very clever. It used three remote-pulse transmitters co-ordinated in such a way as to supply, electronically, latitude and longitude information, by which the navigator could determine the aircraft's position. This is how it received its original name "Trinity", the holy three in this case being three pulse transmitters that constituted the RAF's first radar navigational aid. The system was called the "Goon" box, however, to mask its real name, "Gee". ("G" was for grid, i.e. the electronic grid of latitude and longitude derived from a combination of the three signals received by the aircraft.)

By the time Harris took over Bomber Command, Trinity, by now renamed Gee, was a reality, although it was still in its infancy as a navigational system. With its introduction and use, RAF bombers could pin-point their targets with unerring accuracy. They no longer needed to fly to distant targets in a series of dog-legs, but could fly in a series of curves, always being able to calculate a position relative to the ground, and so deceive German radar operators as to the ultimate destination. Gee was based on radar rather than on radio technology, so that radar, at this point in its development, became an offensive system as opposed to its defensive function under the CH-CHL-GCI-CDU systems.

The introduction of Gee equipment transformed the whole science of air navigation in much the same way that the Knickebein bombing beams helped the Luftwaffe navigators maintain a track to their targets. The advantage of Gee over the Knickebein was the ability it provided for Allied bombers to approach the target along an indirect route, whereas German pilots had to fly the beam. At a range of 250 miles, Gee pin-point navigation, unaffected by wind and weather, was accurate to within 100 yards. From March 1942, when it was introduced, we were able to mount 1000-bomber raids to saturate enemy targets, and to reduce aircraft losses, which, however, still remained heavy.

On the German side, dozens of small Würzburg defensive radar stations were integrated into a co-ordinated network along the coasts and inland for the defence of German territory. Groups of these short-range precision radars controlled the anti-aircraft guns and their associated searchlights, taking a toll of RAF bombers. Würzburg, operating on a wavelength of ½ metre, had a range of about 12 miles. Although this was a short range, the plotting precision of the equipment was excellent, and was therefore ideal for use as a control medium for anti-aircraft guns and searchlights. Equally dangerous for the RAF was the fact that the Germans took a leaf out of our book when they began using night fighters under ground control with the aid of the Würzburg radar.

On the other hand, because of its range limitation, Würzburg was not suitable for the long-range detection of invading bombers. For this purpose, the Germans used the Freya system. By the spring of 1942 we knew a great deal about Freya. It operated on a wavelength of 2½ metres, and tests we conducted in 1941 indicated that Freya was not a precision radar system. We also had knowledge, and a fair amount of intelligence, of the Seetakt, a German naval-gunnery radar that was a precision system. The operating wavelength of this third German system was less than one metre.

Aware of the German radar capability through the findings of intelligence, Churchill pressed for the production of countermeasures, which were not as easy to achieve as it may sound. To take countermeasures, we first had to have more information on the nature of the systems we were up against. Incomplete intelligence on the extent of the networks of the three types of radar complicated the issue of which system should be chosen for the production of jamming devices. The choice was a major decision. The Würzburg could easily be nullified by dropping "window" (clouds of tin foil thrown out of airplanes) to confuse German radar operators, but window was a short-term measure. Similarly, Seetakt could be blinded with the use of window for a short duration. For jamming of longer-wavelength radar like Freya, a device to generate high-power "noise" was needed. Which system to nullify, Seetakt or Freya? That was the question. This is where Air Commodore (later Air Vice-Marshal) Sir Victor Tait, Director-General of RAF Signals and Radar, exerted his influence.

Tait, a Canadian from Winnipeg, Manitoba, had come to the United Kingdom with the Royal Canadian Engineers during the First World War. He was a quiet-spoken, dynamic man with a strong aptitude for logical scientific analysis in his own field. At the end of the war, he transferred from the Canadian Engineers to the Royal Flying Corps and rose to senior rank in the newly formed Royal Air Force. His experience in signals and communications was unparalleled, so he was the natural choice for the position of Director-General of Signals and Radar.

As adviser to Sir Charles (later Lord) Portal, Chief of the RAF, Tait attended Churchill's scientific committee dealing with electronic communications and radar. Like Tizard, he found Lindemann, Churchill's scientific confidant, a difficult man to deal with.

On the subject of German radar developments, Tait identified Freya as the main target for our countermeasures, but he was up against Lindemann, who insisted that Seetakt should be the primary target. The choice of the system on which to concentrate was crucial, for once we committed ourselves, we would show our hand, which would be disastrous if we chose the wrong system. To correctly identify and define the problem is the bugbear of all scientific inquiry, and this was the crux of the continuing dispute between Tait and Lindemann. It was a nice dilemma, which would be resolved either by irrefutable evidence or by the domination of one scientific personality over the other. Both Tait and Lindemann were respected scientists who carried weight. For Tait, it was not a speculative choice. On a large-scale map of Europe, he had plotted every known German radar installation, and the type of station it was.

142

In the autumn of 1941, when the Germans were still using radio to pass their plots back to their filter rooms, eavesdropping by British radio intelligence and scientific tests confirmed that, while Freya was a reasonably successful early-warning system, its precision was limited. Freya radar operators were unable to pinpoint a particular bomber in the night sky with sufficient accuracy to place a night fighter on its tail.

It was this lack of precision in the Freya system, and the accuracy of Seetakt, that convinced Lindemann that Seetakt was the system on which to concentrate. The debate, however, raged in the spring of 1942, by which time, as Tait pointed out, Freya had been updated and was the system most favoured by the Germans to form its main radar-defence network, as was evident from his map of stations and their locations.

To confuse matters further, our "Y" (listening) service, which recorded all German communications, had heard reference to another type of radar, which seemed to be different from both the Freya and the Würzburg radars. This proved to be the Giant Würzburg, which had a maximum range of 40 miles, as opposed to the small Würzburg unit, which had a range limitation of 12 miles. As it turned out, two Giant Würzburgs worked in conjunction with one Freya to form the equivalent of a British GCI station.

As luck would have it, a map stolen by the Belgian underground showed tiny lightning flashes spaced 50 km apart that suggested a line of new defensive radars strung across Belgium, on a main bomber flight path to Germany. In a meeting with Churchill, Tait insisted that the new installations were almost certainly updated Freya radars.

Wachtmeister Christiaan Ganser, the officer in charge of the installation and maintenance of all German radar in northern France, has since confirmed that, starting in the late spring of 1942, he modified all Freya installations to improve their bearing accuracy.

Tait was correct in his conclusion that we should concentrate on jamming Freya radar, and that, therefore, mass-produced jammers should be tuned to the 2½-metre wave-length. Many meetings were held to discuss the subject. Tait had studied the question and he assembled his arguments with persuasive logic. Finally, at a meeting with Churchill in the presence of Lindemann, he expressed his opinion based on a thorough appreciation of German radar, diplomatically suggesting that the facts would have to be verified regarding the system improvements that he believed the German scientists had effected. It was sufficient to convince Churchill. Tait later reported that, in reply, Lindemann said, "So what are you going to do? Send them [the Germans] a postcard?"

However, despite Lindemann, the die was cast. The decision to concentrate on the improved Freya radar was taken, with the proviso that Tait produce irrefutable evidence that the Freya radar system was indeed the German selection for an improved defence warning, and not just an educated guess on our part. Only then would jamming equipment be developed and production allowed to proceed. That decision was to directly involve me in the radar war in a more combative role, although, for the next many weeks, I was blissfully unaware of what fate had in store for me.

12

"JUBILEE"

In 1942, soon after the United States of America declared war on the Axis powers following the December 7, 1941, attack on Pearl Harbor by the Japanese, limited numbers of U.S. servicemen landed once more on England's shores. The Americans were anxious to attack occupied Europe in force, and, through the newly formed Joint Chiefs of Staff Committee, they pressed their British allies to agree. There was mounting pressure from many directions, including that of public opinion, to relieve the eastern front. Soviet Russia was desperately engaged in a struggle with the German invasion of its territory.

The Americans were anxious to do something positive to help the Russians, even if that help were to take the form of a sacrificial raid on the European coast. They were agreed, however, that training, equipment, and a massive build-up of supplies were necessary before a major offensive could be mounted to establish a second front.

Churchill, perhaps with nightmare thoughts of another Dardanelles or Dunkirk on his conscience, was not prepared to make sacrifices. For this reason, a number of impossible invasion plans were dismissed outright by the British Chiefs of Staff. They were, however, willing, even anxious, to conduct an operation in which adequate air and sea support could be provided, so that the survivors of the raid would be picked up and returned to England.

145

Among a number of possible options for mounting a raid was an assault on the port of Dieppe, which offered a number of advantages to the planners. The port was near enough for the RAF and the Royal Navy to provide the necessary offshore support, to give covering fire-power and back-up from the air for the assaulting force, and to give adequate assistance to remove the survivors of the raid by sea. The decision was taken to mount a major assault on Dieppe under the chosen code-name Operation Rutter.

Rutter was to involve the use of 5000 troops. An armoured column of thirty tanks, RAF 11 Group for air support, and commando units would "spike" the heavy coastal guns that protected the eastern and western flanks of Dieppe. A small group of Darby's Rangers of the U.S. Army, which had been under heavy training in commando tactics in Scotland, would have a place in the action alongside British commandos. The main force, however, would be made up of Canadian units, which had been in England for two years without seeing action.

The general plan of the assault would be to silence the coastal batteries protecting Dieppe; at the same time a heavy bombardment of the frontal area of the port, by both sea and air, would precede the landing of tanks, which would clear the way for the infantry. The tanks would then divide into two flanking columns that would converge on the Dieppe docks, from where they would be evacuated with the main assault force.

Principally, then, Rutter would be a Canadian operation, with the support of British commandos and Darby's Rangers, the Royal Navy, and the RAF. For Air Commodore Tait's purpose, Operation Rutter offered an ideal opportunity to investigate a Freya station at close quarters, to determine if the installation had indeed been updated and was what one would now call "precision radar". A Freya station was located on a high cliff overlooking Dieppe.

Freya 28 was located on the cliff-top above Pourville, two miles west of Dieppe, which meant that a force separate from the main assault would attack the village. In any case, the coastal batteries to the west of Dieppe were to be dealt with by the right flanking commando force, Number 4, led by Lord Lovat.

Assuming that the station operated precision radar, Tait wanted firm evidence in the form of hardware. Was it equipped with any anti-jamming devices that the Allies were not aware of? Did it include navigational features associated with ground control of air interception? What was the quality of the German cathode-ray tubes? pulse generator? receiving equipment? It would be helpful to investigate such features, and to compare them with our own. In order to neutralize the Freya system, we had a great deal to learn, which is how I came to be involved in Operation Rutter.

As far back as August 1940 I had expressed my willingness, in writing, to my superiors to offer my services for special missions in which my expertise would be of value. Since we needed to investigate a station that was an intrinsic part of the German night-fighter defence system, it would be necessary to send in a knowledgeable investigator—someone skilled in the newly developed GCI technology. I was, therefore, a natural choice for the task. However, there were two other reasons why I believe I was selected to accompany the raid. First, there was my original report from Rosehearty on a ground-control interception system; secondly, in February 1942 I submitted another lengthy report on radar jamming devices.

In the GCI type of station we had developed, the control of night fighters had become skilful and precise, whereas conventional surveillance radar, such as our Chain Home and Chain Home Low systems, required little or no expertise in the technique of air navigation and night-fighter control. By the late spring of 1942, my own station alone had accounted for about 100 German aircraft operating at night. This is where the report on jamming techniques brought me to the attention of Sir Victor Tait.

From the summer of 1941, when Bold Head received a

147

VHF direction-finder tender for air-sea rescue work, I had been conscious of the assortment of VHF transmissions emanating from the blob of headland known as Cap de la Hague, on the Cherbourg Peninsula. Cap de la Hague, like Bolt Head, projected out to sea. Powerful transmissions from the Freya on the Cherbourg headland could be clearly heard on our VHF 1132A receiver. I had also constructed a simple wide-band receiver to observe the transmitted radar pulses from the station. It was this combination of radar signals and radar pulses, all from the same source, that led me to speculate on the growing importance of Freya, and to wonder what we should do about it.

There was no one in 10 Group in Exeter with whom I was allowed to discuss radar or a proposal to mount a surgical raid on the station to obtain interesting bits of Freya equipment such as the pulse generator, receiver equipment, and signal-filter equipment. This kind of material would tell us how German radar technology worked, its sensitivity, and its vulnerability to interference. At Ashburton, however, just south of Exeter, was the local headquarters of 60 Group, the RAF's technical division that serviced the entire radar warning network in southwest Britain. Here I was able to discuss the subject with Squadron Leader Charles Keir, the senior technical officer at Ashburton. I proposed to him the idea of a raid on the station. Pinprick raids on German-held territory to keep the Germans on their toes and to gather essential intelligence were then fairly common, but I was not aware of the ambitious plan for Operation Rutter at the time.

Keir listened sympathetically to my proposal, and with interest to what I had to say about the advantages that might accrue. As for military operations of the kind I was proposing, he said that they were outside his province, although he did promise to pass on my suggestion to the Air Ministry at Whitehall. Then, on February 27, 1942, a successful attack on the Würzburg radar installation at Bruneval took place.

Bruneval, a village about twelve miles north of Le Havre, was at Cap d'Antifer and was situated on top of a 400-foot-high chalk cliff. The Bruneval raid, conducted by commando units formed from Scottish regiments (the Black Watch, the King's Own Scottish Borderers, and the Seaforth Highlanders), was a spectacular success. Flight Sergeant C. W. H. Cox, a radar mechanic, volunteered to accompany the raid to carry off whatever interesting equipment could be removed, and in this he succeeded admirably. For this mission he was awarded the Military Medal.

The Bruneval haul provided an immense amount of information on the operation of the Würzburg radar, including confirmation of the wavelength on which it operated. We now knew for certain what methods we should use to confuse the Würzburg operators. The whole question of measures needed to counteract German radar was a dicey one, and a decision was not to be taken lightly.

The fact is, given the quality of German and British radar technology, whatever measures one side took would be taken up quickly by the other once the game was given away. In view of the different operating wavelengths of the various types of radar units, the means of jamming had to be carefully considered.

To counteract Freya radar, we would have to use one of the systems we had tested on our own CH system before the war. This was the use of an aircraft-installed transmitter that emitted a mass of confusing pulses, named "railings". Then there was "CW" jamming, where the picture on the radar screen would jump violently up and down. Thirdly, there was the most potent system, the transmission of "hash" or "noise" on the radar wavelength. This device (codenamed Mandrel) could absolutely paralyse the sensitive radar receiver. (It is interesting to note that post-war research has revealed that certain types of moths emit sonar railings to protect themselves from marauding bats.)

During the two years preceding the outbreak of war, the Mazda Company, a branch of British Thomson Houston

Corp., had developed a radar screen that, used in conjunction with coloured optic filters, enabled CH radar operators to see through the unsynchronized hash of a Mandrel-type transmission to observe the signal of an aircraft blip. Given the dogmatic confidence of German radar experts, it was doubtful that the Germans had taken the trouble to develop a screen such as the Mazda Anti-jamming (AJ) cathode-ray screen, for they would see no need for this kind of defensive development work.

In 1942 Mandrel, or white-noise jamming, seemed to be the correct countermeasure to defeat the German radar early-warning system. Nevertheless, the production of hundreds of Mandrel transmitters for use on bombers rested on Tait's decision to target Freya radar, and he had to be sure it would be manufactured for the correct wavelength. Whereas in 1941 we could eavesdrop on German radio-plotting of aircraft and ships, restoration of the European telephone system during German occupation meant that we could no longer listen in, and could not evaluate their radar's performance.

The argument between Tait and Lindemann on whether to target Würzburg or Freya radar (as mentioned, Lindemann had earlier thought Seetakt radar was the main German system) naturally involved the argument on the use of "window" foil versus Mandrel transmitters. Lindemann favoured the short-wave Seetakt, which could be countered by "window"; Tait stuck to the improved Freya system and the Mandrel countermeasure. "Window" was relatively inexpensive to manufacture; Mandrel transmitters were expensive production items. The Bruneval raid gave us the intelligence information we needed on the Würzburg—its precise operating wavelength and frequency; Tait needed to obtain similar intelligence on the Freya system. This is where my report on jamming devices, submitted to Squadron Leader Keir, bore fruit.

In May 1942 I was ordered to report to the Air Ministry, Whitehall, to meet with Sir Victor Tait. Sir Victor, a lean man with a thin face and piercing eyes, was a pleasant fellow and gave me a cordial welcome. Standing rigidly to attention before him, I was overawed by his lofty position and commanding presence. The smell of leather from the easy chairs, the polish and impeccable order of his office, in which the cleared table-top before him bore nothing but an inkstand and a single file, were overwhelming.

"Sit down, Flight," he said casually.

"Yes, sir!" I sat gingerly on the edge of the chair, set to one side of his desk.

"So! You want to go into battle, do you?"

"Not particularly, sir, but it seems to me . . ." and I launched into an explanation of what I'd already written in my report.

He heard me out without interruption, paying attention to what I had to say and only occasionally glancing at the report in the file before him as though to get confirmation that I was following the logic of what I had said there.

"Do you think you could carry out the investigation?"

"I'm sure I could, sir," I said with confidence.

"It's a bit dangerous, you know."

"I'm well aware of the risks, sir," I said.

He tapped the report. "Your data is on the Freya station opposite Bolt Head. I'm more interested in Freya 28 at Dieppe. Would you be prepared to tackle that?"

"Of course, sir," I said readily.

"Well, I think you should think the matter over. What part of London are you from?"

I told him.

"Then I suggest you take the night to think about it and come again in the morning when you've made up your mind."

"Very good, sir," I said. It was a pleasant interview, but I knew it had come to an end. I stood up, put my hat back on my head, saluted, and left.

The next morning I appeared before the Chief again and reaffirmed my intention to take part in the raid. It was a shorter interview this time. The Air Commodore lifted the telephone and, in a quiet voice, informed the Senior Intelligence Officer (SIO) at Combined Operations HQ that I'd agreed to take part in the operation, then directed me to the Senior Intelligence Officer's office for my first briefing.

The briefing was in sharp contrast to the interview I'd had with Sir Victor Tait, and I'm not sure even now that I would have gone through with the mission had I not had that first interview with Sir Victor and given my word. The Senior Intelligence Officer was Wing Commander the Marquis de Casa Maury, to whom, I confess, I took an instant dislike. His manner was supercilious and lofty, and having me stand strictly to attention while he droned on and on in a dry monotone didn't improve my ill-feelings towards him.

Stressing the fact that I must communicate with nobody, even members of my own station, he outlined my part in the forthcoming commando operation.

"Obviously, Nissenthal, you must not be captured," he said. "Ten men will be delegated to help you in every way; if, however"—and his eyebrows contracted with meaning—"you are wounded and you cannot be brought back to England, they will be ordered to shoot you. Also, in your escape kit will be medication that will enable you to depart from the earthly scene in quick time." He paused. "You realize that you are under no obligation to go. If you want to back out of it, now's the time to do so. You can return to your unit and nothing will be said."

It was the damping tone that put my hackles up. I wanted to tell him what he could do with the mission and walk out, but I didn't. Bravado, I suppose. Anger. I held my tongue and stared back at him, keeping my expression neutral.

"I understand the conditions, sir," I said with heavy emphasis on the "sir". "I accept."

The Wing Commander couldn't fail to register the suppressed emotion in my voice. He stared at me for a moment, then slowly lowered his eyes to some documents before him. It was an age before he raised them again. "Nissenthal," he said, accentuating the un-English sound of my surname, "what I can't understand is why a Jew would volunteer for such a dangerous operation as this." He added quickly, "You will, of course, get nothing out of this, you know."

"We're not given to expecting something out of everything we do, sir," I answered, thinking that he could make what he damned well liked out of it.

I must admit I was taken aback. It would be hard to recall any time in my upbringing, my apprenticeship, my work, or my service life when I'd had to put up with the Wing Commander's brand of racism. I'm not saying that I didn't hear anti-Jewish remarks in my youth, but in my circle of friends and acquaintances they were relatively uncommon and rarely directed at me. Perhaps I just moved in the right circles. Still, it was an undisguised insult, and I'm pretty sure he understood my reaction, for he half apologized for his remarks and offered a weak smile, saying, "I wanted to find out if you'd break under the pressure."

It was a hollow sort of excuse to make, and I wondered why he'd gone out of his way to provoke me. Seething with anger, I held my tongue, waiting for him to speak. He finally did, saying at last, "Well, that's settled then, and you'll have to accept the condition of permanent silence on the matter of the death order, if you return safely."

"I do, sir."

"After twenty-five years no one will believe you anyway."

Leaving his office I also began to have second thoughts about the odds against getting back at all, which seemed to increase the more I thought about them.

It began to dawn on me, some time after the interview, that I'd probably done something very stupid in agreeing to the understanding I'd come to with the Senior Intelligence Officer as a condition of going on Operation Rutter. It wasn't as though I was a foot-soldier trained for combat. To volunteer to accompany the raid under a virtual sentence of death was an insane decision, which I was not allowed to discuss with anyone.

Having a few hours to spare, I took the Underground to get home to spend a while with my family. Like most Londoners, my parents had decided to stick it out in the city rather than move to some less dangerous place in the country, but I was always fearful of their being bombed out. I visited them as often as I could, and as they had no idea I was in town I knew my arrival would be a surprise.

During the interview with Sir Victor I had asked for permission to discuss various aspects of the radar side of the raid with a knowledgeable scientist, and he had arranged a meeting with Don Priest, one of the original Bawdsey team. Priest was then with the Royal Establishment at Malvern in Worcestershire, so I soon travelled to Malvern to meet with him.

As Don and I were old friends, we had a pleasant meeting. We discussed the problem in the Wintergardens in Malvern under a sunny sky, to the sound of twittering birds.

"You're quite mad. You know that, don't you, Jack?" he said.

"Someone's got to do it," I said. "It might as well be me."

"Better you than me, then." He grinned.

"Let's supposing you were going to do it, Don. How would you go about getting what the Chief wants?"

"Look at it this way. You have a similar station to a Freya at Bolt Head, so you're probably in a better position to judge than I am, so what would you do?"

"You mean if the Jerries attacked Bolt Head?"

"Yes."

"If someone were to cut our telephone lines when the station was in operation, I guess we'd have to revert to radio communication. Anyone listening to our traffic would learn a great deal." There was no need to elaborate on this to Don Priest. He knew as well as I did that an expert listening in on ground-control interception radio traffic would soon find out how we controlled aircraft from the ground. He would be able to work out the angular sweep and operating arc of the radar, the height, range, and frequency of the signal, and the signal strength and, of course, its limitations.

"Exactly!" said Don. "So that's the first thing to do. Next, if there's a chance of getting inside the station, remove a cathrode-ray tube and any specification documents that will give us an insight into how a Freya operates." He went into more detail as to particular components in which we would be interested—a radar transmitting tube, tuning equipment, and anti-jamming devices—but he also went on to warn me what to expect concerning defence. Since the Bruneval raid, all German radar stations were ringed with masses of barbed wire, mines, and machine-gun emplacements. As at Rosehearty, the German station defences would be bristling like hedgehogs with explosive barbs.

In the weeks following my late-May meeting with Don Priest, I was kept busy. Between carrying out my normal duties at Bolt Head and travelling around the country on briefing sessions, I got my introduction to the Canadian South Saskatchewan Regiment, who were to provide my escort of ten sharpshooters.

With the exception of Privates Leslie Thrussell and Graham Mavor, I was invited to choose my own bodyguard from the ranks of Captain Jack Mather's "A" Company, the company that would be responsible for attacking Freya 28. Riflemen Thrussell and Mavor had been picked for the escort duty by Captain Jack Mather, because he rated them the regiment's two best crack shots, who could be relied on to stick to me like limpets, and to administer the *coup de grâce* if necessary. In fact, Thrussell took a demotion from the rank of corporal in another company to take on the job of escorting me. For the rest, I chose the toughest-looking characters I could find, mostly from the platoon of an Irish-Canadian sergeant, Barney McBride. As it turned out, taking so many men from McBride's platoon was to lead to his feeling animosity towards me, though I didn't think about it at the time. I suppose I should have selected more people from other platoons in the company.

Of the remaining men who made up the escort, I knew them simply by names I made up for them. There was a French Canadian I called Frenchie, a lanky Saskatchewanian I called Lofty, a chain-smoker I called Smokey, and Bud, Silver, Jim, Charlie, and Roy.

I should mention that, even in wartime, it is illegal to order a soldier to be shot without the due process of a court martial, which is the only process by which summary execution is permissible, and only then for very specific crimes such as high treason or desertion to the enemy. There was certainly no other circumstance that would have legally allowed for my execution. I had to concur with the plan and I did, but the thought of it caused me to have many restless nights.

I spent a great deal of time during the next few weeks going over the planned raid with Captain Mather, and the particular part "A" Company of the South Saskatchewan Regiment was to play in it, attacking Freya 28 on the cliffs above the village of Pourville. The basic plan was simple. While the main force made a frontal assault on Dieppe, the South Saskatchewans and I would land a little to the east of Pourville and scale the cliffs to the station.

I had butterflies in my stomach from time to time in the days leading up to embarkation, and the fluttering increased when we actually boarded the transports and waited offshore near Southampton for the weather off the French coast to clear. A Luftwaffe reconnaissance flight that late July evening was followed by an early-morning bombing attack on the assembled convoy. The air raid and the bad weather caused the concellation of Operation Rutter, and the whole unwieldy strike force was disembarked and sent back to its holding areas. For my part, I simply returned to Bolt Head to continue with my night-fighter duties.

The weather was fine, the July sun was warm, and the South Devon coast was an ideal place to be in summer. If we had not been operating a night-fighter control station with Spits roaring down the runway on take-off or returning from Rhubarb patrols, it would have been difficult to imagine a more peaceful spot. Between aircraft take-offs and landings, the lush Devonshire countryside was tranquil and quiet, with the steady hum of bees buzzing about the buttercups and cowslips, birds twittering in the hedgerows, and rabbits hopping in and out of their burrows. I remember the intoxicating solitude of Devon away from the immediate vicinity of Bolt Head Aerodrome, though the sudden angry roar of intruder Focke-Wulf 190s' bombs and cannons frequently interrupted the tranquillity.

The first time they came after my return startled us all out of our complacency. I don't think we ever got used to these frightening raids. The four undetected 190s came in low over the water and appeared above the airfield while our own Typhoon fighters were still on the ground. Even local CHL had its blind spot close to sea level, and the Germans skilfully exploited this defect. Coming in low up the Salcombe Estuary, they clawed up over the headland, dropped bombs on the runway, and roared around the perimeter shooting up any likely-looking targets.

The Typhoons were scrambled and took off without being shot up in their ascent, and gained height. The GCI duty controller provided them with a running commentary on the raiders as though the Typhoon pilots could not see for themselves what was happening. Then, as the first Focke-Wulf roared down the runway and over the rocks that marked the seaward boundary, a waiting Typhoon swooped in like a sharp-eyed kestrel for the kill and the enemy plane plunged into the sea.

Oblivious to his fate, his three companions continued their marauding shoot-up of the headland, strafing parked aircraft under camouflage nets, outlying petrol dumps, and scattered buildings like trigger-happy gun-slingers. High above now, the Typhoons, anxious not to pump their own hardware into the headland, waited for the bandits to get out to sea. As soon as that happened, they attacked, engaging in a sea-level air battle in full view of the station ground crew watching from the headland.

The Luftwaffe pilots were no amateurs. Twisting, turning, performing an amazing display of aerobatics, they slipped away from their attackers and streaked towards France, merging finally into the distant horizon. We heard later, when our fighters returned, that the chase had continued to the French coast, where a second Focke-Wulf was bagged as it was climbing to negotiate the French cliffs.

After this, the Focke-Wulf visits and their lightning attacks became more frequent, for the Germans had obviously pinpointed Bolt Head and other stations in our warning network along that stretch of coast. The attack was typical of the many raids we had from the Luftwaffe fighters at this period. So, between our night work, tracking incoming bombers and directing our Beaufighters, and early-morning visits by enemy fighters, our lives were full of excitement.

I don't want to give the impression that the air activity, either that of the enemy or our own, was continuous. It wasn't. There were hours of inactivity and long stretches of summer to be enjoyed while we waited for the next round, but during those short periods when all hell broke loose, five minutes of being shot at like an insect on the ground felt more like a lifetime. When the firing was over and the contestants had retired to their respective corners, the summer silence was serene and the situation once more normal. This was my life for most of 1942. The Germans were simply emulating the Poles of 317 squadron.

On August 17 I had just come off night duty and was in the office completing the night's report before plodding wearily off to the hotel to sleep when there was a telephone message to say that a Humber Snipe was waiting for me at the guard room. As the driver had no pass, the guard wouldn't permit him to drive into the aerodrome.

Hurrying over to see what this was all about, I found that the car had been sent to take me to London post-haste. After a rushed shave, bath, and breakfast, I clambered into the back seat of the car, and promptly went to sleep, while the driver sped towards Exeter and on to the main road to London. Our first stop was at the Senior Intelligence Officer's office in Whitehall.

"Well, Flight," said Wing Commander Maury. "It's on again."

"I see, sir," I said, but I'd already guessed it was. Operation Rutter had been cancelled; the new name for the same operation was Jubilee, and this is the name by which the Dieppe Raid has since been known.

"And you still don't want to back out? There's time yet, you know. If not, the 'no capture' clause still holds good."

I was getting used to his brusque manner by now, so I said, "That's all right, sir. I've got ten good men to make sure I comply to the letter."

He looked at me with a quizzical expression, unsure what to make of my reply. I thought for a moment that he was going to charge me with insubordination for being flippant. Then he sucked the side of his lower lip, squinted at me, and said, "Very well, off you go."

In addition to my side-bag, webbing, and the rest of my regalia, I carried a small bag of assorted tools: a collapsible hack-saw, wire-cutters, pliers, and other assorted instruments, plus the escape kit with which I'd been issued for Rutter. The kit included a packet of Horlicks tablets, some Benzedrine pills to keep me awake, and a cyanide tablet to put me to sleep for good. I transferred the bag to my left hand, saluted, and left his office. It was on the tip of my tongue to say, "Thanks for the good luck, sir," but I thought it best not to; no point in jeopardizing the chance of a lifetime at this point.

At 1A Richmond Terrace, the HQ of Combined Operations, I had a more cordial reception, a briefing of sorts, and transportation was arranged to Southampton via the village of Crondall, where the Combined Operations Command for Jubliee was in session. Coming down the front steps of Richmond Terrace with my Special Operations Executive (SOE) officer escort, I suddenly saw Quentin Reynolds, the U.S. correspondent. He recognized me from our journey to London during the blitz. He started to speak to me. The SOE lieutenant gripped my arm and warned me not to speak. I gave Reynolds a shrug of apology and walked on. At Crondall Command HQ, a secluded Victorian mansion on the outskirts of the village, I was met by an energetic and fully accoutred Marines major. Once I was ensconced in his tiny office, he announced that there had been a change of plan, and "A" Company would be going in without tank support. He didn't say why. From the evidence of the air photographs, I thought that there'd be little chance of overcoming Freya 28 defences without armoured support. I said so, and the Major agreed.

I was disturbed to hear, moreover, that there would now be no "softening-up" of the forward defence area on the beaches by aerial bombardment because it was thought casualties among the civilian population would be too high. "And good hunting to you, too, mate!" What else could you say? A staff car took me to the King George V Dock at Southampton in the company of the same SOE officer.

The designated vessel for the South Saskatchewan Regiment was the Dutch motor ship *Invicta*, sister ship of the *Princess Beatrix*, to which we had been assigned for Rutter. Now, in the fading light of evening, the men were already on board, crowded on the dock side of the decks. They were making a horrendous din, when the car stopped and I got out, by banging their helmets on the steel hull of the ship. The nearer I got to the gangplank to go on board, the louder and more rhythmic became the banging. These Canadian troops, who had been in training in Britain for the past two and a half years, were like war-horses champing at the bit. Time and again they had been promised action, special operations, or transfer to a fighting front, but always, at the last moment, the orders had been cancelled. On the grey camouflage-striped hull someone had daubed the message "Get us action or get us home". The men were in a surly mood.

Two and a half years of inaction had left a mark on both the Canadian troops and the British troops on home service. Canadians, with larger pay packets and engaging accents, were popular with the women, and took to English pub life as a mudlark takes to the Thames. British troops, on the other hand, their emotions no doubt conditioned by envy, came to regard the inactive Canadians as "the tough Canucks who saw all their action in bed". The Canucks reacted in kind, so that fist-fights and brawling became commonplace, and the deteriorating situation had become serious enough at one time for the authorities to consider returning some of the frustrated soldiers to Canada.

It is easy enough to understand how the Canadians felt. They had enlisted to fight and were anxious to give an account of themselves. This was one of the many reasons why, when it came to a decision as to which troops should undertake Jubilee, the Canadians were chosen.

To them, the arrival of a Staff car signified another pep talk by a visiting officer, and as the car reached the gangplank the booing and jeering reached a crescendo. Slinging my blue sidepack over my shoulder, I stepped out of the car and looked up at the men. The racket gradually subsided. I'd already met the men of "A" Company, of course, so I was no stranger to them. They had christened me "Spook" because of the pale complexion I had acquired from working by night and sleeping by day. The men of "A" Company realized that my being taken on board meant that something must really be in the wind, because the booing changed to a rousing, on-going cheer as I made my way up the gangplank and stepped on board. The gangplank was retracted, and the ship made ready for sea.

13

FREYA 28

Newt Blackeley, the Staff Sergeant radio expert of the South Saskatchewan Regiment (SSR), was the first to greet me. Lieutenant-Colonel Merritt, the battalion's CO, had assigned Blackeley to show me around when I first visited the regiment at its depot. Now we stood together at the handrail chatting over the background noise of a ship being prepared to put to sea.

The *Invicta* slipped its moorings a short time later and moved past the checkered floating forts of the Southampton water to take up its lead sailing position. There was a great deal of noise. The crew ran up a Spanish flag and hurriedly erected a dummy funnel of wood and canvas in case we were observed by a last-minute reconnaissance.

From the receding shore came the muted sound of an air-raid siren. I felt a certain detachment from what might be going on overhead in favour of what was happening closer to the water. The activity on board was intense. On the one hand the crew were busily making things shipshape, winding in the mooring lines, clanking and banging turn-buckles, hawsers, and winch hardware back into their rightful places. On the other, the fighting men were being issued with small arms and ammunition, smoke canisters, grenades, and bangalore torpedoes. The activity was intense. NCOs shouted their orders; officers were scurrying about overseeing the

NCOs; and the men, issued with brand-new arms, were industriously cleaning and oiling them for action.

For security reasons, the men had been told that this was another exercise, but now no one believed the story. They knew that this was the end of their training and that they were about to see action at last. It was August 18, 1942, and there was a festive mood in the air. Some men were singing "Praise the Lord and Pass the Ammunition". It might have seemed that the euphoric atmosphere was too high-pitched to last, and that the noise would die down and the men would then become quiet and reflective, but they didn't. Very few tried to sleep.

The *Invicta* crept out to sea around the Isle of Wight and headed due south, taking its position at the head of the other ships. They were motor vessels, and the powerful throb of straining effort could be heard from other ships in the convoy. We surged south in the darkening night, leaving behind us a track of phosphorescent water.

The only person who did not seem to be in a happy mood was Sergeant Barney McBride of "A" Company. He had been unhappy about the break-up of his platoon when I chose my escort. The previous afternoon, I discovered later, he had pulled his men back into the platoon when I failed to put in an appearance. He was disconcerted when I did arrive and was vociferous in his complaints. Perhaps it was his way of voicing his excitement, but he was damned, he said, if he'd have his platoon's efficiency destroyed by having to escort "a Limey civvy in uniform".

"Barney's as mad as a hatter," one of his men said. "Better keep out of his way, Spook."

"It's nothing," I said. "Excitement, that's all. He'll calm down."

At about 2 a.m. on the nineteenth, having run south and backtracked, the convoy halted ten miles off the French coast. Out there, in the far darkness, were Dieppe and Pourville, our targets. It was chilly. A strong breeze was blowing, and the sea was choppy as we assembled in our allotted positions for boarding the landing-craft. Over the ship's loudspeakers, a disembodied voice instructed us to inflate the Mae West life-jackets we were wearing under our battledress jackets. "Don't over-inflate," we were warned. "Just enough to feel it."

Then it was our turn to scramble over the side and down the nets to the landing-craft faintly visible below.

"Move back, please." "Squeeze in there!" "Hey, fellow! Mind my head!"

Jostling together, making room for more, we arranged ourselves as comfortably as we could for the two-hour journey to the beaches. Compared with my fellows, I was lightly equipped. I carried a revolver in my holster, and my RAF-blue haversack, stuffed with my hand tools, over my shoulder. It would all be so simple, I told myself. We'd been over the plan until we felt we knew it by heart.

"Green Beach", between the village proper and the Freya station, was our designated landing-place. In the half-light, "A" Company would attack the Freya 28 on the heights between Pourville and Dieppe to the east. The other SSR companies would secure a bridgehead, and the Canadian Camerons would then dash a mile or two inland to attack the aerodrome.

Riding the swell, feeling queasy from the movement of the small craft, we left the protection of our mother ship and headed for the shore. It is hard to imagine the thrill of the knowledge that five flotillas were clawing their way through the pitch-black night on their way to their respective beaches. On either side of the *Invicta*, but stretched mostly to the west for a good ten miles, 5000 men, with tanks and armaments, had been disgorged. With V8 engines throbbing, the blunt bows of the vessels of our armada ploughed through the choppy sea. The spray came in regular bursts and we bore it with the patience of men who had more important things on their minds.

On the flanks of the advancing landing-craft were the vessels of No. 3 Commando Group to the east and No. 4 to the west. Their task was to spike the radar-directed coastal guns that defended Dieppe. More or less in the centre of the assault force was the main attack group that would strike Dieppe. As soon as they were able, 30 tanks would land and fan out, some to the east, some to the west. The tanks moving west would strike inland behind Pourville to reach an assembly point to escort the foot-sloggers from Freya 28, Pourville, and the aerodrome back to Dieppe for evacuation.

With my knowledge of British radar, I knew that any force attacking the coast of Britain would be detected at least twenty miles out to sea. This danger had to be accepted, but there was no way I could tell the troops about it—most of them had still never heard of radar. Then suddenly, when we were half-way, judging by the time that had passed, a brilliant flare lit up the eastern sky and I knew this was for real. The Germans, trust them, were on their toes.

For the next ten minutes, a horrible racket went on. I could have sworn that the assault craft stood out like bandit racoons caught in a blaze of light. Miraculously, however, the crump of shell-fire and the booming of heavy automatics accompanied by tracers darting this way and that ceased. We learned after the raid that two German E-boats, escorting a small convoy, had discovered the assault boats of No. 3 Commando, which was practically wiped out in the ensuing exchange. A Polish destroyer, the *Slezak*, however, had silenced the German E-boats.

The approach of the E-boats was detected by CHL Beachy Head, and its warning message was, in fact, received by HMS *Berkeley*. The ship's skipper, following orders, maintained radio silence, because he took it for granted that the command destroyer, HMS *Calpe*, had also received the message and would take action. It didn't, and the loss of No. 3 Commando Group became the first casualty of the action. The incident nearly put an end to our own involvement.

Petty Officer Hobday, the skipper of our landing-craft, shouted out "Do we turn back?" to the packed men. A burst of chatter broke out amongst us, and Barney McBride's voice rose above it in a thunderous parade-square bellow. "Turn back now and we'll take over! We'll go in by ourselves, okay?" We ploughed on.

A rum-filled canteen began making the rounds, being passed from mouth to mouth. Silver, Bud, Les, Lofty, Jim, and the rest of the escort had jostled to be together, getting between me and the front of the landing-craft. I took a swig and handed it to the man behind me. "After you, George," said his companion, and I watched the bottle disappear among the bobbing sea of tin helmets.

Suddenly a massive column of blackness loomed above the forward ramp. Standing towards the back of the craft, near the wheelhouse, I had the impression we were on a collision course and were about to be smashed on the rocks. Then, just as suddenly, the engine bells clanged, echoing back from the cliffs, and the engines slammed into reverse. Our forward motion checked, we moved to starboard and began moving west with the looming cliff on our left. We were moving parallel to the incoming waves, which gave the landing craft a sickening sideways motion.

Under that high cliff, I had the feeling of being spied upon by hidden eyes and expected at any moment to be sprayed by hand grenades and machine-gun fire. However, nothing disturbed the night but the slap and whoosh of water against the side of the craft.

The rectangular armoured tub in which we were being carried rolled like a ponderous pig. A few minutes passed and the dark cliff fell away and disappeared from our line of sight. We did a ninety-degree turn and headed towards the beach.

We hit the solid ground with a thud and the firing began. The ramp crashed down and the men in front ran. Waiting for my turn to run, my revolver drawn for action, I felt a tug on my lanyard and spun around, caught off balance.

166

Somehow I got my revolver planted in the man's stomach. He leapt back quick as a fox, into the "on guard" position with his bayonet to my throat.

"Bastard!" he said, circling me. It was McBride.

It happened so quickly I had no time to react.

"Pick up that bangalore," he commanded.

"Take the bloody thing yourself," I shouted. "I've got my own stuff to carry." At long range, from the cliffs, we were under fire from light artillery. An incoming shell hit the landing-craft and exploded. In the chaos that followed, I ran down the landing-ramp and escaped.

In the dim light, I could see fallen men, some still, some moving painfully in slow motion. I crashed my way up the stony beach for all I was worth, for the protection of the sea-wall, and realized only then that the Royal Navy had planted us slap-bang on the doorstep of the houses occupied by the German garrison. We should have landed east of the village, nearer to Freya 28. Brilliant! Les Thrussell and the rest of the escort were waiting for me, huddled under the sea-wall.

"Took your time, didn't you, Spook?" said Les.

In the gloom of pre-dawn light, we probed the high barbed wire on top of the wall, looking for a position for our narrow scaling-ladders. By sheer luck we came upon a large gap, obviously used by the Germans to give swimmers access to the beach.

The defenders were taken by surprise, so many had joined the fight in their underclothes. There were bodies lying slumped in doorways and over window ledges. The fighting was general and intense, with the rattle of hundreds of small arms punctuated frequently by hand-grenade explosions. The night wind at sea had died down; the stench of cordite was strong, and for the first time I could detect the sickly smell of death.

We moved as a loose group through the village, running, stooping, taking cover from house to house. We worked our way to the east of Pourville and took shelter for a while in the front portal of a tiny white stone church. Just ahead of us, a bridge, heavily defended on both sides by concrete pill-boxes, crossed the small River Scie, which ran into the sea to our left. Beyond the bridge the road snaked steeply up the hill to the cliff a good mile away, on which the Freya station was located.

Across the road, opposite the church, was the Hôtel de la Terrasse, and because it seemed a safer haven than our present position, I dashed across the road. There was a deafening explosion which bowled me over in mid-flight. I lay where I had fallen for a while, the breath knocked out of me and frightened out of my wits. Unhurt, I was a bundle of nerves, and was desperate to relieve myself.

I ran to the nearest house with Bud close by, hurried up the stairs, and went into the lavatory, where, to my relief and comfort, I sat while the battle raged. A small hitch in this admirable state was the lack of paper, which Bud solved by handing over a wad of propaganda leaflets intended for the local inhabitants.

"Here! Use this," he said.

The house was nicely furnished with highly polished table and chairs, a dark wood sideboard, and a padded footstool. There was a brass clock in working order on the mantelpiece above the cast-iron fireplace, flowers in a vase, and small-print curtains at the windows.

The incoming wave of South Saskatchewans had flowed to the bridge, where an enormous concrete road-block barred our passage. From its black slit a stream of bullets spewed out. The pre-dawn darkness had passed, and the sky was getting lighter by the minute. "A" Company, under Captain Murray Osten, was attacking furiously.

While waiting in the house for the rest of "A" Company to deal with the bridge defences, we were joined by Jim, Charlie, and the rest of the escort, all waiting for the next move. Bursting with impatience, Charlie suddenly threw his rifle to Bud and removed two grenades from his belt. Holding off only for a pause in the automatic fire, which signalled that the Germans were fixing another clip of ammunition on their weapon, he sprinted to the pillbox, shoved two grenades through the slit, and was on his way back to us when they exploded.

We very nearly met the same fate ourselves when a party of SSRs prepared to lob grenades into the house. We were saved by Mavor, who happened to be outside when they arrived.

"For Christ's sake! Don't throw that!" he yelled. "They're our blokes inside, not Jerries."

Colonel Merritt arrived and led the way. Oblivious to the fire, he encouraged the men to engage, and, with a large body of men, he led a fierce attack and cleared the bridge.

At one point Captain Osten seemed to have been hit and slithered down the bank of the River Scie, but he picked himself up and led his company forward. The defensive positions further up the hill had the bridge in their gunsights. Crossing the bridge in a hail of bullets, we followed the van of "A" Company fighting its way uphill. The bridge was littered with the bodies of "A" Company men.

Thrussell and the rest of the escort didn't like hanging back while their comrades were taking enemy fire. They joined in the fight, despite my entreaties that we stick together, so we too began taking casualties.

McBride was hit after crossing the bridge. He was some distance ahead when he fell, so I saw him go down. I was in a state of confusion from the noise and pandemonium going on about me, but I found Barney McBride's death, as I thought then, saddening. He was a decent sort of fellow in spite of his animosity towards me. He was fiercely loyal to his men, which I had to admire; his anger at having the best of them taken from his command by a pale-faced Limey was understandable. Both Graham Mavor and Charlie were killed soon after crossing the bridge. Fortunately, Osten found a sheltered lane leading uphill towards the station.

Zigzagging uphill, crawling in the drainage ditch alongside the gravelly, flinty road, we left our dead and wounded strewn behind us. Part way up the hill, one of the men carrying a backpack of mortar shells was hit and blown to pieces by his own shells. It was a stunning explosion that paralysed my ear-drums and silenced the world for a while, for I was within twenty feet of him when it happened.

The machine-gun and sniper fire from higher up the hill was still intense. Over a distance of a mile, and after more than an hour of heavy fighting, "A" Company, which was originally a hundred strong, was reduced to two dozen men. Of my own ten-man escort, there were seven left, of whom three were walking wounded.

At last, we reached the top of the hill with what was left of Osten's company. There was a T-junction in the road at the western perimeter of the Freya station. The remnants of the company lay in a shallow ditch slightly downhill, out of sight of the German defenders. I crouched with Osten. Every time anyone raised his head above the hedgerow there was a burst of machine-gun fire, which slashed the hedge like a murderous scythe. We were in an awkward position, and had no means of securing entry to the station in our reduced state.

"Well! There it is," said Osten, smiling, crouched in the hedgerow by my side. "Take it if you want it."

Freya 28, slightly uphill of our position, was so near and yet so far. The main building, protected by a strong perimeter of barbed wire, was a round, sand-bagged structure with a small radar-hut pod just visible at the base of the pivoted aerial. The first thing I noticed was that the sweep of the aerial was restricted to a 180° arc. That observation suggested to me that, unlike our GCI aerial which could sweep through 360°, the connection between the Freya radar aerial and the operator's hut was co-axial cable. The British stations used a clever electromagnetic rotating coupling which permitted the aerial to be turned through a full circle.

But the other observation was far more important. I noticed that the aerial would frequently stop and swivel repeatedly through a very narrow arc as the operators DF-ed individual targets. Normally, the aerial rotates uniformly through its full 180° arc, picking up its targets only on each successive sweep. The fact that the aerial I was watching would focus in a particular direction with short little sweeps, before moving to a different angle to repeat the movement, showed that the operators could focus at will on individual targets. Beyond doubt, Tait was right: Freya was a precision radar. That was what I had come to find out—but I was 150 miles from home.

When I had discussed the raid with Don Priest at Malvern, nothing about my mission had been cut and dried, despite the very desirable objective of getting at the equipment. Cut the telephone wires, yes. That would force the station operators to use radio to relay our aircraft movements by radio, which our monitors could listen to. If I didn't return, they would at least know the degree of accuracy of the German radar. There were other, equally important, tasks. Don very much wanted an "output" valve from the transmitter, and any anti-jamming equipment I could find. An output valve would give us the main operating characteristics of the equipment: the wave-form amplitude, power output, voltage, and wattage. I was to look for the unusual with an eye based on seven years' experience of radar development. One couldn't tell for sure what might be available simply from careful ground observations—such as my discovery of the coaxial feature of the aerial. If breaking into Freya 28 was the primary objective, the primary question to be answered by so doing was, had Freya 28 been updated since 1941?

With Osten's only radio transmitter out of action, we had no means of calling up reinforcements. We were too few in number, and too lacking in mortars, to overcome Freya 28, and until this situation was changed I had no choice but to become a fighting foot-soldier like any other rifleman in the South Saskatchewans. Our only hope lay in bringing up more men and more mortars. Following discussion, it was agreed that I should try to get help. The guns of the Royal Navy could help to reduce the station if we could make contact. I therefore set off for Pourville with Les Thrussell and Frenchie for company, leaving the five remaining men of the escort with Captain Osten.

Taking advantage of the available cover to avoid German machine-guns and snipers, which were still in action and covering the approaches from the bridge, we dodged and darted like drops of water on a hotplate. The path was littered with dead and wounded, and in the area of the bridge over the River Scie the fallen lay in great numbers. We crossed the bridge one at a time, dashing and dodging zigzag fashion to make ourselves hard targets for German rifles.

Pourville was a shambles. Considering it was still early morning, between eight and nine o'clock, enormous damage had already been done. Shell-torn, burning buildings, scattered fallen masonry, exploding shells, with small arms blazing away, making one hell of a racket, smoke, and clouds of dust made the village an unpleasant place to be. Men were crouched in every conceivable position that provided sanctuary. We made it to the garage where Colonel Merritt had originally set up his headquarters, and learned from one of the wounded men lying there that battalion HQ had been moved to the casino nearer the beach.

At the casino, I asked the radio operator to contact the command destroyer for support fire, but he laughed, saying that was out of the question. There hadn't been any radio contact all morning, he said. The Germans were using our radio wavelength, and had given orders in English to confuse us. There was also static on the line, and snatches of BBC broadcasts filtered through, so radio communication was useless. What chaos!

Pourville was a beleaguered village. From the high ground of the cliffs where Freya 28 stood, concealed German rifles and automatics rattled away at anything that moved below, so it was a matter of moving from refuge to refuge to survive. We desperately needed reinforcements if we were to have any hope of overcoming the station. Going outside into the

172

courtyard at the back of the casino, we saw the wounded sheltering against the wall. It was a sweltering hot day, and a dozen Frenchwomen in sleeveless summer dresses were busily attending the Canadian casualties. They had set up tables with bread, wine, and water and were doing what they could to alleviate the suffering. The women were calm and deliberate about their work, and I was amazed at their courage. They had asked in their naïveté if they could go back to England with us.

With the aid of a Cameron sergeant I regrouped some men who had a mortar, and we decided to get back to "A" Company. We were near the sea-wall when a mortar bomb exploded and knocked us off our feet in a blinding cloud of choking dust. Getting to my feet, I ran coughing and sputtering back along the main street and over the bridge. Then I made my way laboriously back up the hill to Murray Osten and his remnants of "A" Company. Frenchie was a short way behind. The French Canadian was as dogged a soldier as any I'd met. He didn't say much, but he stuck like glue. Reliable, taciturn most of the time, he would look at me sometimes with an expression in his eyes as much as to say, "Okay! What d'you want to do now?"

"Where's Les?" I asked.

Frenchie answered with a shrug. Had Les been wounded or killed by the mortar blast that had knocked us over? I tried to remember when I had last seen him, taking it for granted that he'd stick close to me. The fact was, I was closer to Les than to anyone, for he was the companionable soldier, talkative, ready with a suggestion, advice: "Keep your head down, Spook." "The bastards have you in their sights." "If not them, me," and his square, thrusting jaw would drop in laughter. I found it hard to believe that he was actually dead, and I felt sure he'd turn up sooner or later.

Murray Osten asked if I'd got the message off to HMS *Calpe* and I shouted my answer, explaining that ship-to-shore communications were all but non-existent, as the Forward

Observation Officer could not be found. Osten, powerfully built and self-confident, was a born leader. He listened, then would make up his mind and give orders; his men obeyed.

The situation on the hill was unchanged. The Freya compound, with its barbed-wire fence, its machine-gun nests protected by sandbags, its blast wall for the concrete radar building, and its rotating-antenna hut, was simply impregnable to our small force. There were other defences too: many turfed slit trenches manned by riflemen outside the conclave.

Overhead, the cloudless blue sky formed a backdrop to the dogfight that was raging. The fight was fiercest over Dieppe, a little less than two miles to the east, although the sky in all directions was alive with fighters in action. The black plumes of smoke from downed aircraft rose lazily in the still air from many places, and the distant rat-a-tat-tat of fighter cannons was constant. As many as three parachutes could be seen floating earthward at that moment. In the first three hours of fighting, 160 RAF and 55 Luftwaffe aircraft were knocked from the sky above Dieppe.

While we were impotent and the air battle raged, I watched the Freya aerial in action. From the manner in which the aerial moved—evidently plotting and following individual flights of aircraft—I was satisfied that Freya 28 was a precision installation. I would dearly have liked to get my hands on some of its equipment.

I was getting desperate, but no more so than Murray Osten. A short while later, I made a second trip into the village to get reinforcements. This time I was accompanied by Smokey, who, dark-skinned and swarthy with black hair and penetrating eyes, was taking frequent swigs from a water bottle filled with rum. I had chosen him for his strong face, but he was not the most reliable companion to have with so much liquor in him. On the second trip into the village, I met Colonel Merritt, who had visibly aged in the past couple of hours. His face was pale and drawn, and there were dark shadows under his eyes.

"We need more men up there, sir," I told him. "Otherwise we can't get into the station."

He ordered a corporal and a section of unengaged Camerons to follow me; one man with a mortar, another with the shells. Once again I made my way up the hill to Captain Osten, and I arrived at the top with fewer men than I had started with. We were at a standstill; it was a stalemate, but I was loath to give up trying to do something. Further up, on the crest of the hill, I could see the telephone lines that connected the station with the outside world.

If the lines were cut, as I'd discussed with Don Priest at Malvern, the station operators would have to pass their plots by Morse code on radio transmitter, and our own radio people could listen in. German plotting accuracy could be compared with that of Beachy Head, which was plotting the same tracks. I told Osten that I was going to cut the telephone lines. The men would cover me during my crawl to the telephone pole.

The hedged road in which we sheltered continued around to the rear of the compound, so I began moving uphill along it crocodile-fashion. The earth reverberated from the thud of gunfire and I felt it through my body. The thick grass, long and ripe in the heat of summer, was perfect camouflage, but no protection against the bullets. At the turn in the road on the crest of the hill, I saw the telephone lines where they left the station, forty yards from where my companions lay. The structure that held up the wires consisted of three poles spiked together near the top, two vertical and a slanted one between them, to be used as a "ladder" in emergencies.

Suddenly I realized the reason for the heavy ground vibrations, and I felt my hair stand on end. I had crawled alongside a buried machine-gun post that was pumping lead through a slit almost at ground level. The position faced Pourville, so it had a limited line of fire, either into Pourville or into the beach area. I was now behind the position, and as I moved away from it towards the telephone lines, the ground reverberations lessened.

Safety is relative. The air battle going on was continuous. Enemy and Allied fighters were in deadly combat, twisting and turning in the bright blue sky. As I looked west, down to the Scie Valley, I could see fighters going at it hammer and tongs, almost at ground level. From my vantage point on the hill, it was an amazing sight, because at times the fighters were below me.

I crawled from the ditch and, still crocodile-style, wriggled along the side of the road, protected by a bank on the left. I passed one triple-telephone pole and made for the next one, which had wire leading into the station. From my RAF knapsack I fished among the tools for a pair of wire-cutters and found two pairs: one I slipped into my battledress knee-pocket and the other I put into a top pocket.

It was now or never. As I have described, the structure was made of two poles set a couple of feet apart, with the third one locked between the two at an angle to form a stable ladder. Rising from the grass, I wedged myself between the poles and worked my way to the top. Fifteen feet above ground level, I snipped at the nearest wires. An intense thrill of excitement ran through me as they twanged and fell away. The easiest way to get at the others that were out of reach was to stretch. I clung to one wire for support, and attacked the rest. I was about to cut the last one of the six when a horrendous racket of machine-gun fire unnerved me. Convinced that it was directed at me, I lost my footing in a moment of panic, swung on the line for support, and hung there by one hand. Then I cut that line with my free hand and dropped fifteen feet to the ground with a crash.

I was stunned by the fall and by the gunfire which seemed only a few yards away. I lay huddled on the ground, sure I'd been shot. Even though I was shaking, I tried to collect my thoughts and courage. I tried to gauge where the firing had come from and surveyed the world from my worm's-eye view. Should I stay put or try to get back? If anyone was watching and I kept quite still, they would think I was dead. Perhaps if I moved slowly I wouldn't be noticed. Cautiously

I slithered forward. Nothing happened. I moved again and stopped, then moved again, getting bolder, and soon, snake-like, I reached the ditch behind the gun emplacement where the others were bottled up. In my panic I'd forgotten my tool kit. I was angry about that, but was not prepared to go back to recover it.

It was now past nine o'clock in the morning. I told Murray that the only way to get into the Freya station was with a tank to lead the way. The tanks that were supposed to help the Pourville force leave the battle area were to meet us from the direction of Petit-Appeville, a hamlet a little to the south of our position. I hadn't given up the idea of crashing into Freya 28 and felt that if I could get one tank to help we'd be home free.

Murray Osten saw the sense of this, and said he'd hang on until he heard from me. With this assurance, I set off for Pourville for the third time. With me were Sergeant Roy Hawkins of the SSR's Security Section; Jim, the silent one of the escort; Smokey, still with a supply of rum in his bottle; Lofty, wounded in one arm but game for anything; Silver, prematurely grey, but the oldest and most cautious of the group; Bud, stocky and red-cheeked, always smiling; and Frenchie, on whom I felt I could rely without Les around. I wasn't mistaken. They were all tough men, each one different, but reliable escorts every one of them. On the double we moved down the hill.

We left the Freya perimeter in silence, grim-faced and determined. Crouching, crawling, and scrambling for the protection of every bump in the ground going downhill, we came in sight of the Pourville beach with its wreckage of shelled and burning landing-craft, helmets, and equipment. The beach was littered with bodies, more obvious now that the tide was going out. Two hundred yards of pebble beach now separated the sea-wall from the sea. It would be a long journey for anyone attempting to escape from Pourville by the same route we had come in. I looked forward to meeting up with the tanks.

We recrossed the bridge, the scene of so much carnage, and reached the main street of the village without incident. In more than four hours of bombardment, a lot had happened. Pourville was changed beyond recognition from the neat village it had been: rubble, masonry, broken glass, and the wreckage of smashed door and window frames was littered everywhere. We moved from house to house with caution in ones and twos, crouching, running, pausing, hurrying again, taking advantage of anything that provided a shield, for we were still in sight of German rifles and machine-guns high on the hill. It was hot work. Battledress with a Mae West under the jacket and full equipment was no sort of fighting gear for the height of summer. I paused to exchange my battledress blouse with a lad who no longer needed it; my own was torn to ribbons.

We were dishevelled, dirty, and sweaty, but constantly alert to every rifle crack, every burst of automatic fire and mortar-bomb explosion. Sometimes the firing was sporadic and often it was intense. One never got used to the noise or forgot to be cautious. The next bullet, bomb, or falling piece of shrapnel could be for you.

We paused for a while in silence at the stone church, and sheltered with our backs against its rough stonework. I had a good idea of the geography of the area from maps I'd studied before the operation. I told Roy Hawkins that the crossroads at Petit-Appeville would be the best place to wait for our tanks, and he agreed. Petit-Appeville lay on the main road from Dieppe to Le Havre. The tanks would have to come by that route to reach Pourville.

I press-ganged some more Camerons. "Come on," I said. "Let's get going." We set off, casting glances back the way we had come.

At the end of the street we turned right and hugged every wall as we headed south for Petit-Appeville. "There's a fork in the road further along," I said. "We take the right fork for the crossroads." There was argument. Lofty didn't think
we should venture further south because inland could be

crawling with Germans; Smokey was beyond caring, saying only, "We're with you, Spook."

Conscious of the need to disrupt enemy communications, I told everyone to break any telephone lines we found. I still had a pair of wire-cutters with me and used them whenever I had the opportunity. The men simply ripped away any lines they found strung out on posts.

We were joined by another small group of Canadian Camerons near the end of the village, which increased our number to fifteen. The Camerons were all fresh and ready for a go at the Jerries. They were there all right: to our left on the far side of the Scie Valley and on the high ground above the string of houses. They had good fields of fire, which made Pourville a hotspot for anything that moved.

We waited for a lull in the firing to make a dash across the road for the protection of the next building. We were nearing the end of the houses and it was impossible to tell where the shots were coming from. In the lush valley of the Scie there was a flat, marshy area of bulrushes. Beyond that, the wood-covered slope of the rising hill provided good camouflage for hidden rifles. The houses on the right side of the black-top road were built into the hillside, and their white-painted wooden verandas at the second-floor level gave them their distinctively French look. The ragged hedgerows and clumps of red, blue, and yellow summer flowers heightened one's sense of unreality along the empty road.

Further along, on the left, we came across an orchard, branches of the apple trees laden with plump red fruit hanging over the fence, a soccer field with the goal-posts framing a lighter green patch of mown grass, a patisserie, and, opposite, a deserted petrol station.

"Hey, look!" said someone. "A car."

We ran across the road to investigate, Jim, Roy Hawkins, Frenchie, and I. It was a sleek, black Citroën, just the vehicle we needed to get along to meet the tanks. There was no ignition key inside. Standing to one side, I lifted the knocker on the front door of the house with the foresight of my rifle.

"Oui?" said a mustachioed pale-faced Frenchman, opening the door. He wore a dapper suit, obviously an office worker.

Frenchie explained we needed the car to get to Dieppe. The Frenchman ignored this. He asked Frenchie who we were, and Frenchie told him, "Canadians."

"Canadians?" He said he didn't understand. When Frenchie demanded the car keys and the Cameron corporal butted in with "Don't frig about" in English, the Frenchman held up his hands in surrender, and jabbered something I couldn't understand.

The man refused to co-operate, so one of the men tore the ignition wires out of the car with his bayonet, and we got on our way.

A short time later we left the road and entered the forest. It seemed easy going by the map, but it was a struggle moving among the trees in full equipment. We hoped the trees would conceal us from the Germans on the hill, but we had to move across open spaces to advance, and the German sharpshooters struck two of the Camerons, who were badly wounded. We tended their wounds with field dressings, made them as comfortable as we could, and went on. Eventually we arrived at the Petit-Appeville crossroads (which after the Dieppe Raid was renamed Le Carrefour des Canadiens).

At the crossroads we spread out and sat down on the grass bank to wait, for we were now out of range of enemy fire. Without making ourselves obvious, we had a good view down the Dieppe road in the direction from which we expected the tanks to appear.

The air battle above was still raging, and we watched the Spits and Messerschmitts fighting it out. If I painted the bright blue sky that hot summer's day it would be with fighter aircraft wheeling and turning and black smoke trailing earthward, but how could one paint the sound and the fury of it all? The sounds came and went in waves: the roar of high-powered engines as a pair of fighters come roaring in low up the valley and disappeared over the hill on our left, the urgent burst and boom of cannons, the distant thump of guns.

During quieter moments, there came the sound of a tinkling and scraping from I knew not where. It was an odd sound, unusual and far away. We were stupefied by exhaustion and heat, and the rest was a welcome relief from the day's heavy work.

The rough scraping sound grew louder, like the steady percussion of a symphony, and triumphantly grew into the joyful noise of powerful engines and the metallic rattle of tank tracks on the hard pavement. I sent a Cameron corporal to investigate and make contact. He came back a short time later, moving at a gallop, and we completely mistook his excited shouting.

"The tanks! The tanks are here," we yelled.

We leapt to our feet and hurried up the bank by the roadside. My God! It was not our tanks at all. Coming round the far bend in the road was German Panzer armour, flanked by cycle-mounted infantry. They saw us at the same time we saw them and scrambled to get themselves in fighting order. We turned into the trees and fled pell-mell back cross-country in the direction we had come from.

We took a short-cut through a nearby orchard, and hurried through the trees in the direction of Pourville. A mortar shell burst among the trees and spurred us on. The Jerries hadn't lost much time setting up their weapons.

Frenchie was reaching for an apple from a tree when he was hit under the armpit by a bullet. He fell to his knees in agony and fainted. He came to while Bud was tying a field dressing on his arm as a tourniquet to stem the flow of blood.

"You'll be all right, Frenchie," I told him. "Come on, let's get on." I placed his rosary in his hands, for I had noted the unmistakable grey pallor of death creeping over his face.

Bullets were flying about our heads; a mortar exploded thirty yards away; we pressed deeper into the trees, scram-

181

bling, running, and leaping hell-bent for leather. Another of the men was hit, in the temple. I stopped to help him, but there was not much I could do. His blue eyes, half closed, were seeing their last of the world. A machine-gun opened up from one of the houses that was now on our left and sent us scurrying even further into the timber. Roy Hawkins leaped over a hedge at the side of the road and fell tumbling and rolling down a twenty-foot embankment that he hadn't seen. He picked himself up and ran to catch up with the rest of us.

There seemed no end to our panic-stricken flight. Going through a farmyard we were attacked by angry, cackling, hissing geese and lashed furiously at them with our weapons. To add to the confusion, the sound of roaring aircraft engines intensified. We didn't know that the "Vanquish" order to abandon the raid had been given some time earlier, and that the RAF had sent in squadrons of support aircraft to cover the retreat from the beaches.

We ran and ran and ran till we thought we had no strength left to go on, and all the while we ran we were being picked off by the enemy. I remember Silver being hit in the stomach and collapsing into a ditch with a choking gasp. Blood and saliva were oozing from his mouth and down his chin; I knew he was dying. Lofty had knelt down with me, but there was nothing we could do. Of the original ten-man escort, Silver was the sixth to cop it. Jim was hit in the shoulder while we were trying to get an abandoned radio set working, but his wound wasn't serious.

We were moving down the valley and were within sight of what had once been battalion headquarters when we saw a German patrol on our left, slightly uphill of us. Damn! We dropped into the ditch and waited. What could we do? We were almost out of ammunition. Lofty raised his rifle and fired. One of the patrol fell and the rest fired their automatics, hitting the wall behind which we were hiding, showering us with stone chips. Further down the road, a trio of elderly Frenchmen, one wearing medals gleaming in the sun, were watching the exchange with impassive expressions. Slowly the one wearing the medals moved uphill, getting in the line of fire, which immediately petered out. No one wanted to shoot a civilian; a German officer on the hill must have ordered his men to hold their fire. The old soldier held his course until, within a few feet of where we lay, his eyes met mine. I knew that he knew the predicament we were in, and I loved him for his courage. Using him as a shield, we took our chances by scampering from our hiding-place and hurtling helter-skelter down the road, past the hotel and the church to the Rue de la Mer, and along a back alley to the casino.

The courtyard and the building were crowded with survivors and with the wounded, far more numerous than when I had been there two hours earlier. The Frenchwomen were still tending the wounded and giving what comfort they could. Our group from Petit-Appeville—what was left of us—must have been the last to reach the casino, for the Germans had us surrounded and were pumping everything they had into the building.

At sea—the tide was still out—landing-craft were braving the shells and small-arms fire from the surrounding heights to get close to shore for survivors. The whine of shells, the barrage of heavy explosive and bombs were simply deafening; the smoke was choking. Clouds drifted across our front, obscuring from view large areas of the sea and the harbour. Landing-craft for several hundred yards could be seen in various states of seaworthiness. Some were burning, others spewed banks of black smoke into the air from ruptured fuel tanks.

The casino was holding out, for there was still ammunition, and still manpower capable of keeping the enemy at bay. Without leadership, the able-bodied survivors crouched behind the wall of the alleyway and fired away, using up the magazine clips that were replenished by their wounded comrades. Lofty was hit by a spent bullet that ricocheted from the barrel of his gun. Bud knelt by him during the last moments of his life, wiping his lips.

The defence couldn't last. It was clear we'd have to give in or suffer the consequences of not doing so, and I was getting desperate to make an escape. I thought we could still make it to the beach and out to sea with a little luck. Roy Hawkins, Bud, and Jim agreed. Smokey, drunk as a lord, was beyond reason, and ready to do the job he'd been assigned to do: putting paid to me if there was likelihood of my being captured—and this looked increasingly possible as the situation deteriorated.

For me there could be no surrender. It was now about 11:30 a.m., and our situation was desperate. I decided there and then to take charge, and organized a party of volunteers to rush a nearby road-block, where, I was told, there were two anti-tank rifles with spare barrels and plenty of ammunition. I got the weapons, and even a prisoner, back to the Battalion Headquarters and collapsed exhausted alongside Hawkins in the front room. We sat amongst the rubble on our reversed tin hats and contemplated the situation, while others set up the anti-tank rifles outside the window.

I felt trapped, closed in. We couldn't hold out much longer. Roy thought we could make it to Spain if somehow we could escape from our present situation. I thought his suggestion out of the question. However, I knew if we didn't escape I'd have no alternative to taking my poison pill as promised Casa Maury.

Bloody hell, I thought, waiting for the anti-tank rifles to be fired. This can't be the end after what we've been through. The casino was under steady shell-fire from the Jerries on top of the hill overlooking us. The choking dust made my eyes water and my throat bricky-dry. I kept my mouth open to overcome the deafening effect of the shell-bursts.

Our fellows should have been firing the Boyes anti-tank rifles by now but they weren't. I got up and went out, to find one of the weapons abandoned. It fired a hefty armour-piercing shell intended for fighting tanks, and, because of the kick of its recoil, was almost as deadly to the firer as to the enemy. With the help of Rifleman Bob Kohale of the SSRs (a helper was needed to load it), I pulled the butt of the rifle into my shoulder and took careful aim at the house on the cliff-top from which the Germans were firing at us. When I was certain I had the weapon sights on target, I fired. The effect was immediate and devastating, like an exploding hand grenade.

The Germans stopped firing, so we had either surprised them or hurt someone. Kohale reloaded and I fired again. During the lull that followed, a group of survivors left the casino and ran for the Pourville beach. The Germans, from another vantage point, opened fire again, and the men who were hurrying to escape by running to the water to swim out to sea began falling like rag dolls. The survivors scurried back to the protection of the sea-wall, while the luckier ones, already in the water, waded out to sea.

I had vaguely noticed a destroyer some distance offshore laying a smoke-screen to protect the landing-craft that were moving around nearer in. It must have seen our plight through high-powered binoculars, because it came closer, approaching the cliff-face opposite the German gunnery position that was keeping us under heavy fire. Suddenly the destroyer fired a massive broadside at close range and a large section of the cliff-face simply disintegrated. The destroyer was HMS *Albrighton*. Its broadside was followed by an uncanny silence, which was immediately followed, in turn, by an almighty cheer from the hundreds of men trapped on the beach.

Another wave of men left the sea-wall and rushed head-long for the water and the landing-craft hovering out at sea. That's when I decided that for myself it was now or never.

Making my way inside the shell of the casino, I located Roy and the others of our party and held a quick council.

"We could smash that window and make a break for the sea-wall if we had smoke canisters," I said.

"Okay, I'm with you," said Hawkins.

So we smashed the window frame with our rifle butts and organized help to throw the remaining canisters outside, then we ran for our lives towards the sea-wall: Hawkins, Jim, Bud, and myself. I'm not sure about Smokey. He was being difficult about it, and hung on to my battledress jacket just as I was going through the window. I hurled him away from me with all my might and scrambled through. No one was going to stop me if I could help it. I don't remember whether or not he followed me. I didn't stop to find out, but ran madly through the blinding thick smoke to the sea-wall.

Choking for breath, I came into the clear air behind the coils of wire that lined the top of the wall. Then I began a frantic search for an opening through which to drop onto the pebble beach. Where the others had got to I couldn't tell. There was too much noise and confusion to be concerned for anything but my own survival at this point. I finally found a gap in the wire and leaped the ten feet onto the beach, landing squarely between two stretcher cases lying against the stonework.

"Try that again," said one of them. "You might land on me next time."

"Sorry, mate. I'm in a hurry," I said—or something like that. I stopped, out of breath, and ruffled his copper hair. He was a young fellow like me, barely out of his teens. He seemed cheerful enough, but he was rolled and strapped into a casualty stretcher and couldn't move.

There had been about fifteen in our escape group when we began our dash for the sea. The Germans in the houses overlooking the beach opened fire with everything they had. Only four of us made it to within a stone's throw of the water. The rest was a haze as I thundered across the pebbles, my boots slipping and sliding over them. Somehow I got ahead of the others.

On the run, I had stripped off my battledress top. I had forgotten that my well-washed RAF shirt, which at one time had been blue, was now almost white, and even with the Mae West it made me a perfect target for German rifles. Roy shouted for me to put my top on again. Then I dropped one of my hand grenades and stopped to pick it up. Roy stopped with me.

"What's the trouble?" he said.

"I lost a grenade."

"Hell! You don't need that now."

I recovered it all the same, and clipped it back on my belt. I was aching in every joint. My muscles were stiff from so much running, my brain was in a whirl. I'd never felt so hunted in all my life. There was shouting behind us. Germans on the sea-wall fifty yards away were yelling—for us to give ourselves up, I suppose. Roy and I were alone. The other two escorts who had stuck with us to the sea had fallen.

"Run!" Hawkins bellowed.

We ran into the ebb-tide, discarding helmets and equipment as we went, over the slippery, shining pebbles that hadn't yet dried in the sun, and splashed through the incoming waves, knee-deep, waist-deep, chest-deep in the rising water, and plunged at last into the sea. The salty water burned our raw, singed faces. I hadn't had a drink all morning and was very thirsty. Bullets were splashing all around us. I shouted to Roy to act as if he'd been hit and to swim underwater. He did, and I did likewise. Fear, I suppose, helped me swim further underwater than I'd ever swum before or since. When I surfaced, with lungs bursting for air, it was to find Roy Hawkins a few yards away. We swam and swam, each looking occasionally to make sure the other was still afloat. Further out, the smoke seemed miles away.

I could see in the distance a landing-craft appearing from, and disappearing back into, the billowing clouds of smoke drifting across the water. Every time it appeared, a shell would come screaming above our heads and burst with spectacular effect off-target. The plume of water seemed to touch the sky.

With my Mae West partly inflated, I found it possible to stop swimming and take a rest. I was becoming increasingly tired, and the rests were becoming progressively longer. I was also swallowing a lot of sea water, but after a while I felt too tired to worry about it. During one of my longest rests, I was hit on the head with a painful and resounding crack. The landing-craft had hit me. A Cockney sailor hauled me aboard and threw the hand grenades over the side, saying, "You won't be needing these any more, mate."

I lay on the waterlogged deck and, with sea water swilling around, tried to tell my Cockney saviour to pick Roy up. Each time I tried, water spouted from my mouth, but there was no sound. Finally I squawked, "Pick up my mate!" My Cockney twang was instantly recognized. With the natural good humour and instant repartee of the East End of London, he said, "What do you think this is, mate? The number eight bus?" Roy was dragged aboard and we lay on the deck like two landed fish, completely exhausted and quite unable to move.

We were fortunate. The landing-craft that picked us up had been having engine trouble. Assault landing-craft needed two engines for steering, and, because one engine was out of action, and while the conked-out diesel was being coaxed into life by the crew, the rescue craft was forced to move around in circles. After a considerable time the second engine came to life, and, to the steady hum of the two throbbing V8 engines, we moved north out of the smoke and were promptly attacked by waiting fighter aircraft.

For me this was the most frightening episode in the whole raid. The petty officer who was the skipper of the craft was holding a Lewis gun to his shoulder, swearing and firing at the German aircraft as he took evasive action. This was accomplished by shouting to the helmsman, "Ninety degrees port," or "Ninety degrees starboard," as two German aircraft came in to attack. The wily pilots scotched his plan by breaking up and coming in from different directions, scoring violently noisy hits on the armoured walls of the landing-craft with their cannons. The racket was frightening, like that of a great sledge-hammer beating on the metal sides of the assault landing-craft.

The ramp was partly down and we were taking on water as we made our way slowly towards the *Albrighton* while the two fighters did their best to finish us off. We dragged ourselves to the stern of the craft to avoid the water. I was looking over the side at the aircraft when I was surprised by the sudden appearance of big black balls of smoke near their wing-tips. The *Albrighton*'s ack-ack guns were firing at them. As we approached the destroyer, the German pilots decided that the accuracy of the Marine gunners was too high, and they sheered off.

After what seemed an age we came alongside the low after-deck of the destroyer, and the landing-craft, once it had lost its forward motion, started to sink. We were quite unable to help ourselves, but the sailors pulled us aboard and we lay on the blazing-hot armoured deck in the brilliant sunshine, feeling that at long last we were home and dry. But even now this was not to be the case. We lay amongst the guns, and watched and marvelled at the marksmanship of the Marine gunners, who were in continuous action against the aircraft which seemed so intent on our destruction.

We were attacked by low-flying, single-engine fighters and by high-flying Junkers 88s who carried out steady bombing runs, which were interrupted by the gunnery of the Marines. Bombs fell either short or ahead of the destroyer, which wheeled around taking evasive action. The worst attack came from a Junkers 88 that came in just above sea level and dropped two bombs, one of which did not go off; the second one seemed to lift the destroyer out of the water. The waterspout from the bomb blacked out the sun and we were deluged by the sea water. There was a large air fight going on overhead, where Spitfires tried to prevent German bomber aircraft from sinking us. The destroyer *Berkeley* had already been sunk by the Luftwaffe, and they were obviously hoping to repeat the success.

The *Albrighton* was the rearguard for the remainder of the small craft that were now headed back to England. A Spitfire pilot with black smoke streaming from the aircraft flipped over at low altitude, jettisoned his cockpit canopy, and came out on his parachute. The *Albrighton* and another vessel headed for the airman, but we were beaten to the punch. Accompanied by good-natured boos from the men on our craft, the other vessel picked up the floating pilot. It was late afternoon, but the sky was still full of aircraft in combat. For the first time, however, there were far more German aircraft than RAF. We slowly made our way across the Channel and arrived at Newhaven late in the evening.

It was after midnight when Roy and I got ashore, thoroughly exhausted. We stepped into a darkened warehouse building, dropped to the floor, and fell fast asleep, only to be roused a bit later by some military police poking us in the ribs. We were transported in a Canadian army truck to a debriefing session at the Canadian Field Headquarters in Sussex. Roy spoke up for me, because as the lone man with an English accent among the Canadians, I was suspected of being a spy who had infiltrated and come across with the returning troops. It was some time before the Canadian interrogation officers accepted the fact that I was British and not a spy. I was given permission and a warrant to return to London on the early-morning workers' train. The ultimate irony was that I could not tell of my mission. There would have been no point in telling these men that I had carried out a successful intelligence raid on a radar station— my companions had never heard of "radar" and would not have believed that it could "magically" detect airplanes at 100 miles. We had tried to get into a radio station and failed—I wrote my "report" and signed it.

As I stood on the platform with commuters who were reading their morning newspapers about the raid, I felt conspicuous. I was a sorry sight. My battledress trousers were torn, I was wearing a Royal Marines battledress top, I could not put on my tin hat as it was out of shape, and my face felt very sore. My scruffy appearance probably explained why my fellow travellers kept their distance, cautiously glancing at me now and then. The small Southern Railway train came puffing into the station, and I climbed aboard. Dirty, dishevelled, and unshaven, I viewed the morning commuters with bleary eyes.

At Waterloo Station I risked arrest by the military police because I was without a paybook or other means of identification. Reaching the ticket barrier, I noted that the M.P.s were talking under the station clock, so I dodged out of their sight and made a dive for the Underground washrooms. An airman was shaving himself and I asked permission to use his razor. He refused. I then looked in the mirror and realized why. My face was dirty and raw from shrapnel and exposure. I cleaned myself up as best as I could and made my way to Westminster Underground station, then out along Whitehall to the Air Ministry building in King Charles Street. The guards at a table in the entrance foyer were difficult and demanded identification, but I told them to telephone Air Commodore Tait. They phoned, verified that I was *bona fide*, and then escorted me through the building to his office. The Air Commodore organized tea and sandwiches before questioning me on the raid. It was a pleasant and welcome homecoming.

14

DIEPPE POST-MORTEM—
THE TIDE BUILDS

The bloody beaches at Pourville—and also at Dieppe from all subsequent accounts—made one thing clear: the Allies would never get back onto the European mainland without much better navigational aids, equipment, and organization than those that went into the Dieppe Raid.

After my meeting with Sir Victor Tait, who congratulated me on my safe return, I was interviewed by Wing Commander de Casa Maury, the Senior Intelligence Officer. Even though the only practical achievement of the mission was that I'd cut the telephone lines connecting the Freya 28 station with the outside world, the information gained was of inestimable value.

Severing the lines, as I have mentioned, forced the German radar operators to use plain-language radio-plotting to enable their air controllers to direct the Luftwaffe fighters. This meant that RAF eavesdroppers could form an accurate picture of Luftwaffe aircraft control, and assess the accuracy, range, and operational efficiency of the Freya 28 station. Comparison of aircraft plots between Freya 28 and the Beachy Head CHL indicated that there was no longer any doubt about one vital fact: the Freya-type radar had been upgraded and was now the main early-warning radar system being used in the defence of Hitler's Europe. It was perfect radar intelligence for our side, and confirmed Victor Tait's conviction that he was correct in identifying the Freya system as the target for any countermeasures we had to take. His radar scientists were now able to develop the electronic strategy that Dieppe had shown to be necessary, for we knew the precise performance characteristics of the Freya equipment, its range, its blind spots, its operating frequency, and its limitations. His conviction justified, Tait issued instructions for the mass production of Mandrel 120-MHz white-noise jamming equipment.

As part of the post-mortem study, I had to submit a report of my experience, with observations and recommendations, which I did. Three things stood out as absolutely necessary to be addressed and resolved: navigation, deception, and improved communications.

The near-grounding of our landing-craft on the rocks beneath the cliffs of Pourville, and our eventual landing right on the doorstep of the garrison defending the village, were object lessons not to be forgotten. With the hundreds of ships and tens of thousands of troops that would be involved in the return to the European mainland, there was no room for this kind of error. We had not been landed on the beach east of the River Scie as planned, and as a consequence the South Saskatchewans had to force a passage across the defended bridge. Their slaughter was a tragedy that had to be seen to be understood. In any future landings, I argued, it would be necessary to effect a landfall with pin-point accuracy. After all, what was the point of gathering intelligence of enemy positions if the assault force was landed at disadvantageous points? The Dieppe experience demanded a navigational system that would allow mother ships and landing-craft to find their positions in the inky pre-dawn darkness. It would have to be a passive system that did not depend on primary radar, which was too easily detected. The need for navigational accuracy was obviously the first imperative, and, as it happened, there was already a system for ensuring an accurate landing location.

As early as 1936, radar scientist R. J. Dippy, one of the Bawdsey men, had suggested a system for the blind approach to a target. Using two radio transmitters emitting synchronized pulses equidistant from a chosen flight path, a navigator could direct a pilot along a given course. Bob Dippy's idea was both logical and simple, but the navigator, although he was able to stay on course, was unable to fix his position along that path. The addition of a third ground transmitter to add this positional data was introduced in 1941. Trinity, as this system came to be known, came into operational use in March 1942. It was as accurate a means of fixing an aircraft's position as any system yet devised, and was used by Bomber Command for the great raids on Germany from 1942 on.

In my report I suggested a navigation system for surface craft and assault craft based on the Trinity (or Gee) system. Ships fitted with Gee receivers could put assault troops ashore within a stone's throw of their intended targets.

The next major obstacle to be overcome was devising some means of masking the approach of any assault force to its target area. I recalled the skirmish that broke out on the left flank of the flotilla during its approach to the French coast, a good five miles from the beaches. Of course, I wasn't aware that German escort vessels had only accidentally discovered the commando landing-craft; on the contrary, I was convinced that our approach had been detected by German coastal radar—a not unlikely explanation for the gunfire exchange. Something was definitely needed to avoid enemy radar detection in the event of an Allied invasion of Europe.

I have explained earlier some of the problems of radar operations and how easy it was to cause "hash" and "white noise" on the radar screen. There was also "window" available. Window drops had to be used sparingly, for foil dropped from a height affected the radar screen only as long as it took to fall to the ground. Besides, once used, it told the enemy one thing: something was being hidden, and a deception revealed is no deception to an intelligent observer. In addition, though window was good for jamming short-wave radar, it was less effective against long-wave transmissions. We needed something better.

The development of the Mandrel jamming device was the answer to this problem. The Mandrel emitted a high-power signal which blanked out a whole sector of the radar coverage. It could be switched on and off at will, and the intensity of its signal varied. A series of Mandrels transmitting signals of a given intensity could provide a defensive curtain behind which probing radar signals could not penetrate. Since the Mandrel signals were set to provide a curtain at the extreme range of German radar, the German operator would have clear "visibility" up to the chosen range. However, any attempt to "see" past the curtain would result in a "muddy" screen.

To test the newly developed Mandrel, a Lancaster bomber, fitted with a handmade unit, was sent on a scheduled air raid on the town of Mannheim, the middle of a line of Würzburg radars called Himmelbett boxes, which depended for their early warning of attack on Freya early-warning stations. Over Mannheim the embryo Mandrel jamming equipment was switched on. It caused confusion among the German ground controllers. The single Mandrel had an effect on almost 200 miles of radar cover. The equipment was then switched off and the test bomber flew deeper into Germany, turned on a reciprocal course, and retraced its track. When the plane was over Mannheim again, the transmitter was switched on, once more with the same results. There was no doubt that this was a central piece of equipment in future plans.

By October 1942, large numbers of Mandrel transmitters were ready and a mass-production program was begun. Two major raids in which the Mandrel device was used were those on Hamburg later in the war and on Peenemunde, the site of Germany's V1- and V2-rocket-testing facility. The net result of our new-found ability to deceive German radar operators was that we could now plan a variety of spoofs, which of course is what happened, but not in such numbers that the Germans might have caught on to the secret.

The third major electronic discipline to be improved was that of communications. The single VHF link that went from Uxbridge via Beachy Head to the Headquarter ship *Calpe* at Dieppe was eventually replaced by the Number 10 set, a microwave pulse-code modulation system, which is now in service around the world. During the Dieppe Raid—and certainly in the Pourville sector—radio communication had been all but non-existent. It had been impossible to contact the destroyers for support fire, to exchange messages between units in the field, or to call up reinforcements. In my report I stressed the need for communication links to which the enemy would be denied access. During the raid, it soon became obvious that the wily Germans, butting in on our operational frequencies, were countermanding the orders of unit commanders, issuing their own, and generally causing chaos.

The pulse-code microwave communication system, developed in response to this need by General J. T. Robertson, one of Lord Louis Mountbatten's early students in radio communications, has not changed much to this day. Paraboloids, those huge reflecting mirrors that today dot the landscape, would be placed on each succeeding hilltop as the Allied armies advanced across France and Germany following the D-Day landings. The system would, for the first time, allow a narrow, unjammable beam of multi-channel communications to be maintained, the use of teleprinters, and the rapid turn-around of decoded Enigma information that could be placed quickly in the hands of the Allied field commanders. In my opinion, the microwave communications system would not have been developed in time for the D-Day invasion had it not been for the impetus given by the Dieppe Raid. In fact, it was not made operational until a month or two before D-Day. I like to think that my report contributed to the improvement of these three important areas of electronic strategy: navigation, electronic deception, and communications.

Radar scientists had developed an entire range of navigational, defence-detection, and deception systems. Using new ideas, variations and modifications of earlier ideas, every device and radar development stemmed from the pioneering work begun at Bawdsey. The available equipment and systems were brought to perfection as a result of the hard lessons learned from the Dieppe Raid. My own involvement in further development of jamming equipment was a direct result of being on the raid.

Convinced that the Freya radar was the system to neutralize, Sir Victor Tait sketched out his ideas for a jamming system, and, because I had come to know him as a result of frequent meetings and technical discussions, the task of making the first jamming Mandrel fell to me, and I produced the prototype unit at the Bolt Head Station.

I was not directly involved in the immediate preparations for D-Day on June 6, 1944, because, in March 1943, I took part in Operation Fortitude in the Middle East. Following submission of my Dieppe post-mortem report and work on the prototype unit, Sir Victor Tait asked me what I'd like to do next. I said I'd like to serve overseas and he promised me, in that case, a holiday in the sun. The holiday turned out to be a job setting up a defensive radar system in the Middle East, using a mobile Ground Control Interception (GCI) system of the type I had helped develop at Bolt Head.

In preparation for Operation Fortitude, I left Bolt Head in January 1943, and moved to Rennscombe Down near Swanage, where the next four weeks were spent training seven GCI crews. In cold winter weather, we moved a complete GCI unit and crew to Liverpool to embark for the Middle East. Two months after leaving port, we arrived at Port Twefik at the southern end of the Suez Canal by way of the Cape of Good Hope.

A short distance from Port Said, at the Mediterranean

197

entrance to the canal, a small airstrip was constructed at El Gamil, and the radar station was set up for business, very much as it would have been at Bolt Head. The Luftwaffe was very active over the Canal Zone, using Crete as its base of operations. For the next few months, the work was every bit as heavy as it had been on the south coast of Devon, for the Desert War in North Africa was at its height, and the Luftwaffe was intent on blocking the Suez Canal. The Allies wanted desperately to keep it open.

There were numerous warlike activities taking place in the Mediterranean theatre of operations at this time, and the remnants of defeated European armies flocked to join the Allied armies in Egypt. Yugoslavians donned RAF uniforms and, to distinguish themselves from the rest of us, wore blue forage caps with a scarlet five-pointed Communist star. We had an influx of Greek airmen from the Royal Hellenic Air Force, as well as veteran air- and ground-crew survivors from the French, Polish, Dutch, and Belgian air forces. We also had a number of sunburnt "civilian" gentlemen from neutral Turkey to train, on the understanding that Turkey would enter the war. Had Turkey come in on the Allied side, I might have travelled more extensively through that part of the eastern Mediterranean, because I had trained enough civilian airmen of the Turkish Air Force to warrant my being sent into the country. As it was, I spent the next two years in North Africa instead.

By late 1943 Allied fortunes in the war had changed dramatically from those in 1940, when Britain and the Commonwealth stood alone. The Allies now had overwhelming strength in men and equipment, determination, morale, and economic superiority. The contention that radar was the pivotal reason for Allied superiority over the Germans has firm foundation in fact. Whether we examine the U-boat war, the war in the air, or increasing successes on the battlefield, radar technology stands out as the common denominator.

Although detached from the centre of radar developments from March 1943 on, I feel qualified to summarize the prep-

198

aration and events that culminated in the invasion of Europe on the sixth of June. My knowledge of the science almost from the time of its inception, my continuing research (which eventuated in the island-hopping mobile radar Type 6000 units), and my subsequent discussions with colleagues in the radar community, are sufficient reasons for me to speak with authority on the subject. There is a pattern and a sequence of events that, in the opinions of knowledgeable persons, gave the British a commanding lead over the Germans in the radar war, despite German technological superiority, ingenuity, and luck.

From 1939 on, the production of special devices and equipment on both sides was simply astonishing. Intelligence and counter-intelligence produced an amazing catalogue of achievements and accomplishments. The Germans had their Würzburg, Freya, and Seetakt, and were working on even newer systems: the Mammoth, Wassermann, and Jagdschloss. They also had the Knickebein transmitters, night-fighter control stations, jamming devices, searchlight and fire-control systems, and U-boat navigational aids, to mention a few of their accomplishments in radar development.

On our side, we had our radar arsenal. The acronyms of devices and systems sprouted like asparagus on a warm spring day. I've mentioned a number in this account of the radar war: CH, CHL, GCI, PPI, IFF, ASV, and Gee. We were also developing a newer system later called OBOE. We had also by this time developed the magnetron microwave generator, and had introduced the use of tinfoil "window" and the radar-jamming Mandrel. In addition, we had accurately identified and plotted the location of hundreds of enemy radar installations.

From August 1942, when the Dieppe Raid took place, Victor Tait, as head of RAF 60 Group Communications, built up an encyclopedic knowledge of the German radio and radar organization. He already had in his Stanmore headquarters a linear British military grid map covering the entire United Kingdom, on which the identity and location of every one of our installations was plotted. On a European extension of the U.K. grid map, German stations were pinpointed whenever they were identified. By the beginning of 1944, almost a thousand sites and installations had been located. Of these, more than seventy precision stations were strung out along the coast of Europe as part of the main German defensive-warning screen. Tait's main preoccupation from August 1943 on was to develop deception devices and systems that would confuse the enemy on D-Day. He succeeded magnificently.

The basic methods of misleading German radar operators were really straightforward. Dieppe had shown us that the Freya-type radar, working on a low-range frequency, was the enemy's main precision early-warning system. We knew they had numerous airborne anti-surface radars working on a shorter wavelength and that they could detect ships as far away as the English coast. We, on the other hand, had the Mandrel jamming device, with which we could generate white noise on their screens at will. There was also window, which, if used sparingly, could add to the confusion. The trick was to know when and how much jamming to apply.

The Germans, of course, were as aware of jamming devices as we were. And, as in the case of the use of nerve gas or other unconventional weapons, any bungled attempt to jam the enemy's radar would provoke retaliatory jamming. The decision not to jam the enemy's electronic eyes and ears was observed by both sides for a long time before it was reversed and limited jamming began. In 1942, General Martini, the German radar chief, used land-based equipment to jam the Allied radar system during the siege of Malta. However, the Malta radar system kept transmitting throughout the German jamming operations, which, we reasoned, deceived the enemy into believing their equipment was ineffective. Only when we had developed and perfected the Mandrel jamming device did Tait realize he had a considerable advantage over the Germans, because their radar screens were flooded with white noise, which completely blanked out radar images.

Use of the Mandrel by our side was sporadic until the preparatory build-up for D-Day began. One such occasion was the great raid on the U-boat factories at Hamburg starting on July 24, 1943. The Hamburg raid was notable for several innovations. Window was used extensively, along with the Mandrel, and Blumlein's computerized blind-bombing system using the magnetron-equipped H2S radar. The magnetron, it will be recalled, was the microwave generator invented by the two University of Birmingham lecturers which the Tizard Committee took with it to the United States in 1940. The Plan Position Indicator (PPI) on the H2S equipment enabled the aircraft's bomb-aimer to see every detail of Hamburg and its approaches on a calibrated radar map.

The sprawling port city was well defended by a concentration of anti-aircraft guns, which were controlled and coordinated with the equally efficient Himmelbett nightfighter control system evolved by Luftwaffe general Josef Kammhuber. The raid was a far cry from the Luftwaffe blitz on London in 1940. During the 1940 blitz there were no worthwhile air defences, and German air crews had it very much their own way. In contrast, the opposing forces in the Hamburg raid were more evenly matched: well-equipped aircraft on the one side, a well-coordinated defensive system on the other.

Hundreds of RAF bombers took to the night sky from airfields all over Great Britain on July 24. Thirty aircraft equipped with Mandrel jammers led the strike force, and the first indication of trouble that German radar operators had was the sudden appearance of hash and white noise on whole quadrants their radar screens. This phenomenon occurred while the RAF bombers were still crossing the North Sea. The interference occurred only when the Freya aerials were turned to the west.

Somewhere in the hash seen on the screens there could

be dozens of bombers, but the Freya early-warning radar simply could not see them. It was as though a huge curtain had been draped across the night sky. With the Freya units immobilized, it seemed that the short-range Würzburg radars would have to carry out the interception, which was only possible within a 40-mile range. Usually, Freya radar plotted individual bombers at a range of at least 70 miles, handing their plots to Würzburg pairs in a carefully rehearsed and practised procedure.

In the Himmelbett system, one Würzburg watched the target bomber while the other watched the night fighter. The fighter controller then arranged the interception. This system depended on early warning by Freya radar to allow the Würzburgs time to focus on the target bomber and to place the night fighter firmly on the bomber's tail.

Mandrel jamming did not affect the Würzburgs. Special receivers on the RAF bombers, however, indicated when the Würzburgs were looking at them, and for the first time dozens of bombers ejected bales of tinfoil strips. To the German operators, it was as though the bombers were reproducing themselves, because, for every one aircraft, there were now a dozen in the sky. In no time a whole quadrant of the sky was masked by what looked like thousands of aircraft signals. The Würzburgs were blinded. Because of this, gun-control radar could not differentiate between window and aircraft.

Luftwaffe night fighters were already in the air in anticipation of the attack, but the pilots were deafened by the racket on their radio control channels, generated by tinsel radio-jamming. The interference made any form of ground control impossible.

The bombers, finding no opposition, flew in low. The first wave dropped incendiary bombs, which set factories, warehouses, and homes on fire. Hamburg had a well-organized fire-fighting service, which set to work coming to grips with the conflagration. The next wave of bombers dropped tons of high-explosive bombs, and the following wave dropped incendiaries. This pattern was repeated again and again for the next six days and nights: by the RAF at night, by the U.S. Air Force by day. The British had served a bitter apprenticeship in the art of bombing and fire-razing during the London blitz and had learned their lessons well.

The razing of Hamburg was one of the many horror stories of the Second World War, taking its place with Dresden, Hiroshima, and Nagasaki. The Hamburg raid was one of the few occasions on which radar jamming was used on a large scale before the Allied invasion of Europe. It could not be repeated often, because, in the same way that we took countermeasures, the Germans would certainly have soon found answers and methods to counter the new jamming equipment.

15

D-DAY—JUNE 6, 1944

The Normandy invasion was, and remains, the largest sea-borne operation ever mounted, and began the process of liberating the European continent. Entire books have been written about just this phase of the war. I will not recapitulate the very complex history of that event; instead, as throughout this narrative, I will highlight only the crucial role that radar technology played in making that massive armada succeed.

For reasons already made plain, Tait restricted the use of spoofing devices in favour of a concerted effort to mislead the enemy as part of the build-up for D-Day. He found, for instance, that Freya radar frequencies were altered to avoid Mandrel jamming, but he discovered a way to extend the unit's operating coverage beyond the range of frequencies at which Freya radar could operate. He also developed a strategy to ensure success.

First, for some months before D-Day, a fleet of vintage bombers equipped with high-powered jammers began a continuous patrol of the English coast to produce a "curtain" and prevent German radar observations of shipping movements. German coastal radar stations were bombed and a careful radar watch was maintained to see how long German technicians took to get the stations back on the air.

Next, dummy convoys of imaginary ships (which were just small launches trailing large, radar-reflecting balloons) were sent intermittently through the channel as feints. Dambuster Squadron No. 617 (which had attacked German dams with devastating effect), adept at low-flying operations, flew in tight circles around the balloon-towing vessels as they approached the French coast, and dropped window, which deceived the enemy into believing that an armada was approaching. Selected radar stations were not destroyed, but allowed to remain in service, so that they could "see" what we wanted them to see.

The deception worked.

Again, surface craft equipped with Eureka beacons could be located by aircraft fitted with Rebecca equipment. The Eureka beacon transmitter was a particularly polite piece of radar. It did not respond with a signal until it was triggered to do so by a Rebecca transmitter. This meant that Eureka-equipped ships did not betray themselves by making unnecessary radio transmissions.

As a final back-up, the H2S radar on Allied aircraft was unaffected by Mandrel jamming, which permitted aircraft equipped with H2S to see where they stood in relation to other aircraft and ships that were carrying out spoofing operations. By allowing some German radar stations a modicum of serviceability, Tait could permit spoof convoys to be reported to the German control centres. The schemes for confusing the Germans proliferated in the tense months leading up to D-Day.

The important thing about German radar as far as we were concerned was that the Germans should be able to use their radar equipment effectively in the immediate vicinity of the French coast, despite the fact that the whole of the south and south-west coasts of England was hidden behind a curtain of Mandrel jamming. The only noticeable difference in the performance of German radar was that the general noise level became a trifle higher than it had been over the previous months, but only when the aerials faced the English coast. This was the only thing that might suggest to them that something was afoot. However, we hoped they would put this down to some natural phenomenon beyond their immediate control.

In discussion with German radar specialists after the war, it was apparent that the extra "noise", a kind of radar interference that, in part, was natural to all radar equipment, was taken for granted. Its variation in strength was not even noticed by most German operators because it in no way interfered with their ability to plot coastal shipping out to the most exteme ranges. This was normally about 35 miles, which was considered sufficient to ensure early warning of an invasion.

During the last week of May 1944, the weather deteriorated. Boisterous, buffeting winds, heavy seas, squalls, and rolling grey-black clouds made life for the German coastal watchers noisy, tiring, and uncomfortable. The weather pattern was a boon to Tait's deception plans. RAF Bomber Command helped make life more miserable by carrying out frequent sorties on German coastal radar installations. It is to the credit of German engineers that badly damaged installations were soon back in service despite successive, massive, and swift daylight rocket and bomb attacks.

On the 70 radar sites lining the coast, there were more than 200 radar installations of various types. Given ingenuity, good repair servicing, and luck, the Germans reasoned that an invasion armada of the size needed would stand little chance of getting close enough to shore to land an army and the equipment needed to sustain it.

Somehow the Germans managed to keep their radar defences alive, but, as a back-up, General Martini, Tait's opposite number, had a second line of radar defence. If every ground radar failed, he had a special force of ASV radar-equipped aircraft on standby for instant dispatch, to investigate reported shipping plots that might be considered dangerous.

The first massive and co-ordinated use of widely located Mandrel transmitters came as a preparatory measure for the D-Day invasion of Europe on June 6, 1944.

The curtain was first switched on one month before D-Day. A long Mandrel curtain of low intensity was spread to accustom the enemy operators to the idea of remote white noise. Gradually the power of the curtain was increased and projected nearer to the French coast. By D-Day, 200 vessels in the invasion fleet were equipped with Mandrel, although their purpose was not known to the ships' officers and crews.

A week before D-Day, the Allied Director General of Signals ordered the aircraft maintaining the Mandrel curtain to move it further out into the English Channel. The aircraft used were Stirling bombers of the RAF and veteran B17s of the U.S. Air Force. German operators observed an increase in the ambient radar noise level, but there is no evidence that they noticed the worsening conditions. Fortunately, the aircraft holding up the curtain were too far away to be detected by German radar. The general reduction in overall performance was accepted because it still did not affect German operator ability to plot their own coastal shipping and convoys. So far so good; the deception was working.

On the evening of June 5, 1944, the Allied armies under the supreme command of General Dwight D. Eisenhower were embarked on the vessels of the largest armada the world had ever seen, poised for the long-promised invasion of Europe. On board ships, at aerodromes, and in forward assembly areas, 1,000,000 armed men, women, and support staffs waited for the order that would launch the assault.

The weather had been deteriorating for days. Gales lashed the coasts, rain squalls struck inland, and there was low visibility on land and on sea. To the staff officers and commanders gathered at Eisenhower's Combined Operations Headquarters, it seemed that the fates were against them. The best-laid plans would go awry unless the weather eased up enough to allow the invasion fleet to put to sea. Meteorological experts from Stanmore and the British Air Ministry kept the Supreme Commander supplied with half-hourly meteorological reports while the invasion force, kept in ignorance of what the decision might be, waited impatiently.

With the exception of the radar community and most senior members of Eisenhower's staff, no one had an inkling of the extraordinary radar measures that had gone into making the invasion possible. Indeed, without the radar developments on which we had all worked so hard, the invasion of continental Europe would have been a rash undertaking.

The weather continued to be foul: squalls, gales, slashing rain, and black storm-clouds. What was perfectly miserable weather for the one side was welcomed by the other, for while the Overlord (code name for the operation) clock was ticking, the weather made the German watchers feel sure that there would be no invasion. In the meantime, with inexorable intent, the curtain of hash was moved closer to the French coast.

On June 4, unknown to the German watchers, two "X"-type submarines, which had a vital part to play in the invasion plan, took up their positions to mark the beaches where the British and Canadian troops were to land. They lay in the shallows a few yards below the waterline and waited, submerged.

The "X" craft were navigational markers, equipped with Eureka transmitters, to provide incoming craft with positive position indicators. In each submerged vessel were three men, occupying space normally used by two men. They were cramped, uncomfortable, and unable to rise to the surface to replenish their air supply. The decision to deploy marker submarines was a direct result of the Dieppe experience, which I had dwelt on in my post-operation report to Victor Tait. The Americans, despite the lessons of Dieppe, deemed this navigational precaution unnecessary, and paid the price in lost lives in their own invasion area.

Using retractable 11-foot masts equipped with the polite Eureka transmitters, the miniature submarines marked the exact landing-spots on which the invading troops would touch land. Their orders were to surface at 3 a.m. on June 5 when they received the signal that the invasion ships had sailed. The Eureka beacon would be used in conjunction with a shielded seaward-looking light-beacon navigational aid. As events transpired, the bad weather delayed the invasion order. The submariners, true to their sense of duty (and orders), remained at their stations for another twenty-four hours. Theirs was an onerous task, and to those brave men the Allied invaders owed an enormous debt of gratitude.

While the "X"-craft crews waited, so did Eisenhower, although his wait was in the relative comfort of the Overlord Command Headquarters. As meteorological conditions in northern France were invariably determined by Atlantic weather, the RAF regularly flew into the West Atlantic to gather the data needed by the pundits who prepared the weather maps. On June 5, Group Captain Stagg, who headed the meteorological team, reported a "hole" in the weather pattern that was moving in fast from the Atlantic. He predicted the possibility of a few hours of less inclement weather, less turbulent winds—but there was no doubt in his report that the heavy weather would return.

General Eisenhower, after much deliberation, had a fateful decision to make. If he was wrong, the invasion would be aborted by the return of foul weather before the troops could be got ashore. In typical American style, and with no dramatics, his order to his commanders was short and terse: "Okay, let's go."

When the invasion fleet put to sea and the ships' captains broke open their sealed orders, their first instruction was to switch on the "position finding" equipment that had been installed in each vessel. Within a short time, more than two hundred high-powered shipboard versions of the RAF's Mandrel transmitters went into service and the entire fleet was cloaked in an invisible shield. It is astonishing that, for years after D-Day, many senior commanders vested with responsibilities in planning and launching the invasion knew nothing of the purpose of the Mandrel switch.

By 3 a.m. (on June 6) most of the tired German operators would have noted that the noise level had increased to the extent that it seemed the main front-end receiver tube was malfunctioning. At this time in the morning, they would probably decide to wait till the day-time mechanic came on duty to fix the set. Already, however, the landing-craft of the British and American fleets would be preparing to land.

Thanks to meticulous planning, the secrecy maintained by Tait's staff, and the conditioning of the German radar early-warning-system operators, the Allied invasion fleet approached the Normandy coast undetected. The masterful plan of deception had worked. The first inkling the Germans had of the invasion of Europe was in the dawn of June 6, 1944, when, during first light, the sea was seen to be filled to the far horizon with the ships of the Allied armada.

AFTERWORD

In the book *My Life in Court* (1958), Louis Nizer, a noted American lawyer, discussed a famous libel case in which he was counsel for the plaintiff. In 1955, his client, Quentin Reynolds, had sued Westbrook Pegler, a syndicated columnist, and the Hearst newspaper empire for libel. It was alleged that, in his column, Pegler had mounted and maintained a libellous campaign in which he asserted that Reynolds had been an "armchair" war correspondent. In one article, he said that Reynolds had covered the Dieppe Raid from the safety of a battleship. The vendetta so damaged Reynolds's reputation that he lost the means to earn a livelihood; his commissions for magazine articles dried up. Because of this, Reynolds sued both Pegler and Pegler's employer for libel.

In an account of the Dieppe Raid in his book *Dress Rehearsal* (1943), Reynolds told the story of Jack Nissen's exploit. He was not aware, however, of Nissen's identity, and had unknowingly used a pseudonym, Professor Wendall, which had been given in the press release issued to war correspondents soon after the Dieppe Raid took place. The release was issued by the Public Relations Department of Lord Louis Mountbatten's Combined Operations Headquarters. Reynolds had in fact been present on HMS *Calpe* throughout the raid.

At the trial, he admitted that he had not met Professor Wendall in person. He had met him, of course, in the person of Jack Nissen, but was not aware of his true identity, even during the trial. Because Nissen was still bound by the British Official Secrets Act in 1956, he was not able to testify to the truth of the Professor Wendall story. Reynolds did, however, win the court case, and was awarded $200,000 damages against Pegler and the Hearst newspaper empire.

Several accounts of Jack Nissen's Dieppe exploits have been published since *Dress Rehearsal*. One of them, *Green Beach* (1975) by James Leasor, was entirely devoted to the Freya 28 incident. In other works, both accurate and inaccurate accounts of the episode and of Jack's radar exploits have been written. These books include *Israeli Secret Service* (1979) by Richard Deacon; *Dieppe: The Shame and the Glory* (1962) by Terry Robertson; *Reynolds* (1964) by Quentin Reynolds; *Die Radar Schlacht* (German, 1977); *De Laatste Landvoogd* (Dutch, 1977); *The Jubilee of Death* (1984) by Raymond Sauster; *Operation Freya* (German, 1975); *A Man Called Intrepid* (1975) by William Stephenson; *Most Secret War* (1978) by R.V. Jones; and, most recently, *Camp X* (1986) by Dave Stafford.

Many of the stories told have been blatantly untrue. For instance, in *A Man Called Intrepid*, which deals principally with the wartime activities of Sir William Stephenson, Nissen is said to have trained at Camp X, a training establishment for espionage agents located near Oshawa, Ontario, and to have been a colleague of Menachim Begin, and other Israeli agents. Nissen's talents and accomplishments in the field of radar are many, but espionage was not one of them.

For a man whose contribution to the development of British radar was considerable, Jack Nissen is relatively unknown. His Dieppe exploit has been noted, of course, because the incident has the appeal of a strong element of courage and audacity, producing a dramatic narrative. At the same time, Nissen's involvement in the early development of radar was far more important than his achievement in the

Freya 28 incident. With respect to the contribution to radar science, his fate has been shared by most of his colleagues from the Bawdsey days.

Watson-Watt, head of the Bawdsey scientists, and the acknowledged father of British radar, is relatively well known, yet even he for some years went unrecognized by his fellow countrymen for the service he had rendered. The enthusiastic reception given to him in the United States shortly after Pearl Harbor, when the Americans sought and received his help, was in sharp contrast to his treatment in England once the war ended. At that time there were many conflicting claims as to who had invented what in radar science, and to settle the matter there was an official inquiry to decide if, in fact, Watson-Watt had been responsible for the development of British radar. The findings of the inquiry under the chairmanship of Lord Chief Justice Cohen in 1952 were inconclusive, although £90,000 sterling was awarded to the group. Watson-Watt was also given a knighthood.

It is a sad reflection on the pioneering work of Watson-Watt and his colleagues that he and they went unrecognized for so long after the war ended—almost seven years. It was because of what the British radar scientists, as a group, considered to be their shoddy treatment when the time came for their contribution to the war effort to be recognized that most of them left Britain between 1945 and 1955.

Watson-Watt moved first to the United States and then to Canada; Bowen and Butement went to Australia; Priest emigrated to the United States; and Nissen went to South Africa, and then Canada. Cockcroft became the head of the Harwell Research Establishment in 1942. Some remained in Britain to live out the rest of their lives in obscurity. Arnie Wilkins, who was Watson-Watt's assistant and who, though mentioned only in passing in this narrative, made a great contribution to radar development, was among those who chose to remain in England. Alan Blumlein, the brilliant physicist who invented so many things we now take for granted, was killed just as he reached the pinnacle of success,

213

when his test aircraft crashed in Wales. He was flight-testing his H2S radar, the radar that was to defeat the U-boats during the spring of 1943.

Watson-Watt often told the story of being caught for exceeding the speed limit during a journey from Toronto to Port Hope, Ontario. Handing him a ticket, the policeman said that he had been caught in a radar trap. Mrs. Watson-Watt said, "Good heavens! My husband invented radar." Indulgently, the constable smiled and said, "That's what the man said about the guillotine when they led him to the execution block."

Jack Nissen made his home in Johannesburg and spent a period training Allied radar operators and air crews on their way to the Korean War. Later, he set up his own electronics business and travelled widely in Europe and North America. On one of his visits to post-war Germany he met General Willy Weber, who, at the time of the Dieppe Raid in 1942, was the officer in charge of Freya 28. The former enemies became lasting friends. Weber reported that, after the raid, one of his men had found Jack Nissen's tool kit and the Avometer that Jack's father had given him in his youth. The German general was unable to say what became of the kit, but its discovery did give rise to speculation among the German radar scientists that the Canadians had attacked Freya 28 for the express purpose of learning its secrets.

In Johannesburg, Nissen also met former German Wachtmeister Christiaan Ganser, the officer in charge of the installation and maintenance of all German radar stations in northern France. Ganser stated that, starting in the late spring of 1942, he had modified all German Freya radars to improve their bearing accuracy, thus confirming that Air Commodore Tait had been correct in identifying Freya radar as the system to be blinded with radar-jamming devices.

During a radio interview in Saskatoon some years ago, Jack was contacted by a listener, who turned out to be the former Sergeant Barney McBride. Contrary to Nissen's belief that he had been killed on the approach to Freya 28, McBride had survived his wounds. The two men bore one another no animosity and in fact became good friends in their new country.

Leslie Thrussell also survived. He returned to civilian life in his home town of Moose Jaw, Saskatchewan, after spending the rest of the war as a prisoner of the Germans. Thrussell was unaware of Nissen's identity for twenty-five years after the Dieppe Raid. He frequently told the story of the British radar scientist whom he had orders to kill if he was in danger of being captured, but was never believed. Then, at the 25th-anniversary reunion of the raid on Dieppe, the Canadian rifleman saw Nissen in the company of Lord Louis Mountbatten, who also attended all Dieppe veteran reunions. Thrussell asked Nissen to confirm to his former comrades that he was indeed the man that he, Thrussell, had been ordered to shoot.

Nissen honoured the Official Secrets Act for the statutory twenty-five years by not discussing in public the part he played in the radar war. In writing this personal account of the radar war, he felt that the time had come to tell his story.

In 1977, opposed to the apartheid policies of the South African government, Nissen emigrated to Canada with his family, to join his mentor, Sir Robert Watson-Watt.

Jack Nissen is still an active man. He is the president of Museum Electronics Inc., Thornhill, Ontario, which manufactures electronic display equipment for museums and science centres around the world. He owns one of the original magnetrons that were manufactured by hand.

Nissen remembers his colleagues as true comrades in the exciting and sometimes dangerous way of life that made up the radar war. It is a fitting end to that period of scientific achievement that the best of both German and British radar developments were eventually combined to provide the world with its modern navigational and communications systems.

A. W. Cockerill

215

Index

(including Abbreviations and Foreign Terms)

A Man Called Intrepid (1975), 212

AFS (Air Fighter Sector), 114

AFS (Auxiliary Fire Service), 86

AI (Airborne Interception), 82, 103, 105, 131–3

Air Ministry, Whitehall, 148, 150, 191

Air Ministry Experimental Station, Westhampnett, 54, 102, 103

Airey, Joe, 7–8, 12

Alberta, University of, 100

Aldeburgh, 10

Allied Director General of Signals, 204, 207

Architectural Faculty of London University, 40

Arsene, 43

ASV (Air/Surface Vessel system), 22, 49

ASV-equipped aircraft, 205

Avo Minor electrical-test instrument, 3, 214

B17s, 207

Bader, Wing Commander Douglas, 93

Baldwin, Stanley, 10

bandits, 122–8, 132–4

Banting, Sir Frederick, 100

Banting, Dr. W., 100

Barking Creek, Battle of, 50–1

Battle of Britain 38, 78, 82–3, 93–4, 119

Bawdsey Manor, 1, 2, 7, 12–16, 20, 23, 25, 27, 35, 41, 53, 139, 154, 213

Baxter, Tommy, 14

BBC (British Broadcasting Corporation), 10, 19, 20, 38, 172

Beachy Head, 22, 165, 175, 192, 196

Beaufighter, 105, 119, 131, 134, 136–9, 161

AI-equipped, 131, 158

BEF (British Expeditionary Force), 70, 78

Begin, Menachim, 212

Bentley Priory, Stanmore, 42

Biggin Hill, 89

binder, 48

bird-cage, 48

Birmingham, University of, 81, 201

Black Watch, The, 149

Blackeley, Staff Sergeant Newt, 162

217

Blackett, Professor P.M.S., 9
Bletchley Park, 44, 80
blimps, 33, 37
blind approach, 30
Blumlein, Alan, 3–4, 80, 83,
    133, 201, 213
Blumlein modulator, 83, 133
boffin, 45
Bolt Head, 107, 109, 114–36,
    197–8
bombing beams, 31
bombing season, 105
Boot, H.A., 82–4
Bowen, Dr. E., 8, 14, 22–3, 83,
    95, 99, 213
Boyles, R.W., 100
Bristol Aviation Company, 105
British Chiefs of Staff, 145
British Official Secrets Act, 212,
    215
Brown, Bob, 5, 26, 52
Brown, Flt./Lt. John, 135–6,
    137–9
Bruneval, 148
Buchan, Mrs. Chrissie, 58, 77
"Bud", 156, 166, 168, 177,
    181, 183–4, 186
Budden, A., 14
Butement, Dr. W.A., 14, 20–1,
    28, 31, 48, 213

calibrated differential gain
    control, 103–4
Cameron Highlanders, 164, 169,
    174–5, 178–81
Camp X (1986), 212
Canadian Field Headquarters,
    193
Carter, Dr. Robert (Bob), 8, 14,
    23
Casa Maury, Wing Commander
    the Marquis de, 152, 184, 192
cathode-ray tubes, 10, 21, 131,
    147, 155

cavity magnetron, 81–2
CD (Coastal Defence), 20
CDU (Coastal Defence U-boat)
    system, 28, 31, 48, 49, 140
CH (Chain Home), 20, 23, 28,
    37–8, 103, 135, 140, 147,
    148, 199
CH Stations, 42, 49, 51, 57,
    135
    Canewden, 47, 50
    Dover, 88–9
    Dungeness, 89
    Foreness, 49, 89
    Pevensey, 89
    Poling, 89
    Schoolhill, 71
    Ventnor, 89
    West Prawle, 108, 136
    Whitstable, 87
Chamberlain, Sir Neville, x, 28
"Charlie", 156, 168–9
Chiswick radio club, 4, 14
CHL (Chain Home Low), 28,
    56, 97, 103, 135, 140, 147,
    199
CHL Stations, 31, 63, 88, 135,
    140, 192
    Foreness, 49, 63
    Rosehearty, 55–6, 64, 155
Churchill, Sir Winston, 10, 11,
    15, 21, 82, 91, 94–5, 139,
    141–2, 145–6
Cockcroft, Dr. John, 48–9, 53,
    56–7, 213
Codes and Cyphers
    Headquarters, 44
Cohen, Lord Chief Justice, 213
Colchester, 5–6
Combined Operations HQ, 152,
    159, 207, 209
Command HQ, Crondall, 160
Cottage Hotel, 111, 113, 116–19
"Countess of Luxembourg",
    123

Cox, Flight Sergeant C.W.H., 149

Craig, Colin, 104

Craig computer, 104

CRDF (Cathode Ray Direction Finding), 13

Crown and Castle, 12

Croydon, 89

Croydon International Airport, 12

Cunningham, Flt./Lt. John, 137–8

Czechoslovakian Air Force, 68

D-Day, 37, 196, 197, 200–1, 204–5, 207

Daily Telegraph, 88

Dambuster Squadron No. 617, 205

Darby's Rangers, 146

Dardanelles, 145

De Laatste Landvoogd (Dutch, 1977), 212

Deacon, Richard, 212

dead reckoning, 14, 139

Derringer, Colonel, 96, 99

Die Radar Schlacht (German, 1977), 212

Dieppe, 146, 147, 151, 156, 169, 163–5, 174, 179–81, 192–3, 196, 197–200, 208, 212, 214–15

Dieppe: The Shame and the Glory (1962), 212

Dippy, Robert, 140, 194

Dornier, 31

double "L" sweep, 101

Dover Castle, 6

Dowding, Lord, A.O.C. (Air Officer Commanding), 50, 83, 119

Dresden, 206

Dress Rehearsal (1943), 211, 212

Duchess of Richmond, 95

Dunkirk, 70, 78, 145

Duxford Wing, 93

Dyce, 64, 65, 67–8, 76

Eagle Day, 88

Eisenhower, General Dwight D., 207

EMI (Electric Music Industries), Hayes, 4, 7, 16, 25

Enigma, 44, 79, 196

Eureka beacons, 205, 209

Field, Dr. G.S., 100

Fighter Command Headquarters, 41, 50, 61, 64, 67–8, 90–3, 130

FIU (Fighter Interceptor Unit), 102, 105

Focke-Wulf 190, 157–8

Forbes, Dame Trefusis, Chief Signals Officer, WAAF, 57

Forbes-Smith, Lance Corporal, 60

"Frenchie", 156, 173, 179, 181

Freya, 32–3, 37–8, 46, 48, 84, 88, 124–5, 141–51, 154–5, 164, 170–2, 174–7, 192, 195, 197, 199–202, 204, 210, 213–15

Friese-Green, Peter, 5, 25, 43, 52

Ganser, Wachtmeister Christiaan, 143, 214

GCI (Ground Control Interception), 65, 82, 102–7, 114–15, 118, 131–8, 140, 143, 147, 157, 197, 199

GEE, 91, 140–1, 194, 199

Gema Company, 31

German E-boats, 165, 194

German V1, 195

GL (gun-laying), 21, 96, 99, 103

GL.IIc, 99, 103, 106
Glycol, 128
Goebbels, Dr., Hitler's Minister
  of Propaganda, 20
Goering, Reichsmarschall, 63,
  80, 90
Goodeve, Lieutenant Neville,
  101
Gordon Highlanders, 54–6, 59
GPO (General Post Office), 8,
  19, 69
"Green Beach", 164
Green Beach (1975), 212
Gregory, Air Commodore, 45
Gyro gunsight, 95

H2S, 201
Halifax, 95, 100–1
Halliday, Bob, 110, 111–12, 117
Hamburg, 195, 201–3
Hamilton-Smythe, Lieutenant,
  125–6
Hanbury-Brown, Bob, 22, 26
Harris, Air Marshal A.T.,
  139–41
Harris, Anne, 3
hash, 27, 149, 194, 207
Hawkinge, 89
Hawkins, Field Security Sergeant
  Roy, 156, 177–9, 182, 184–
  90
Hearst newspapers, 211–12
Heinkel, 85
Heinkel 100, 39
Heinkel 111, 39, 70, 138
Herd, J.T., 14
"Herr Meyer", 90
Herzogin, Cecile, 116
High-fidelity music, 4, 83
Hill, Professor A.V., 9
Himmelbett night-fighter control
  system, 201
Hindenburg, 33
Hiroshima, 203

Hitler, Adolf, 10, 15, 30, 39,
  41, 43, 93
HMS Albrighton, 185, 189–90
HMS Berkeley, 165, 190
HMS Calpe, 165, 173, 196, 211
HMS Courageous, 22
HMS Southampton, 22
Hobday, Petty Officer, 166
Hope Cove, 109–12, 116, 107,
  113, 132
Hornchurch, 89
Hôtel de la Terrasse, 168
Huff-Duff, 13
Hunt, Flying Officer Thomas, 80
Hurricane, 70, 92, 95

IFF (Identification Friend or
  Foe), 23–4, 46, 51–2, 97, 136
Invicta, 162–4
ionosphere, 10
Israeli Secret Service (1979), 212

"Jim", 156, 166, 168, 177,
  179, 182, 184, 186
Joint Chiefs of Staff Committee,
  145
Jones, Dr. R.V., 9, 46, 79, 212
Jubilee, 159
Jubilee of Death, The (1984), 212
Junkers, 31
Junkers 88, 62–4, 89, 189

Kammhuber, General Josef, 201
Keir, Squadron Leader Charles,
  148
Kelly, Dr. Mervyn, 97
Kenley, 89
Kidbrook, 89
King, Mackenzie, 95
King George V Dock,
  Southampton, 160
King George VI, 50, 135–7
King's Own Scottish Borderers,
  The, 149

Kirkwall, Orkney Islands, 83
klystron, 97
Knickebein beams, 30–1, 39, 79, 82, 140, 199
Kohale, Bob, 185

Lancasters, 195
Le Carrefour des Canadiens, 180
Leasor, James, 212
lecher bar, 81
Leipzig University, 9
Lindemann, Professor F.A. (later Lord Cherwell), 9, 11, 142–4, 150
"Lofty", 156, 166, 177–8, 182–3
Lorenz beam system, 30
Lorenz Company, 30
Lovat, Lord, 147
Lufthansa, 30–1, 32
Luftwaffe, 30, 33, 47, 62–4, 67–8, 70, 79, 89–93, 96, 123–4, 127, 140, 156, 158, 174, 192, 201–2
Lympne, 89
Lysander, 28

McBride, Sergeant Barney, 155, 163, 166, 167, 169, 214
McGill University, 98
Mackenzie, Dr. C.J., 96, 98
MacKenzie, William, 65, 68
McNaughton, General Andrew (Andy), 14
Mae West life-jackets, 178, 187–8
Magnetron, 80–2, 96, 199, 201
Maguire, M., 3
Malan, "Sailor", 134
Malborough, 107–9, 112
Mallory, Air Vice-Marshal Leigh, Chief of Fighter Command, 92–3, 130
Mandrel, 149–50, 195, 199, 204–6, 207

Mandrel 120 MHz, 193
Manitoba, University of, 98
Mannheim, 195
Mansford Technical Secondary School, London, 2
Manston, 86, 89
Mark IV AI, 131
Martini, General Wolfgang, 33–5, 37, 88, 200
Martlesham Heath aerodrome, 22
Massachusetts Institute of Technology, 99
Mather, Captain Jack, 155–6
Mavor, Private Graham, 155, 169
Mazda Anti-jamming device (AJ), 27, 150
Merritt, Lt.-Col. Cecil, Battalion Commander, 162, 169, 174
Messerschmitt 109, 85
Messerschmitt 110, 39
Milch, Erhard, 30, 34
"mine map", 49
Mk.II Spitfire, 119–20
molybdenum, 82
Morecambe, 46
Most Secret War (1978), 9, 46, 212
Mountbatten, Lord Louis, 196, 211, 215
MRU (Mobile Radio Unit), 89, 108, 199
Munich Agreement, 27, 28
Museum Electronics Inc., Thornhill, Ontario, 215
My Life in Court (1958), 211

Nagasaki, 203
Nipkow (or Nipkov) disc, 3
Nissen huts, 120
Nizer, Louis, 211
No. 3 Commando group, 165
No. 4 Commando group, 165
"noise", 27–8, 142, 208

Normandy invasion, 204, 207, 210
North Foreland, 49
North Weald, 89
NRC (National Research Council of Canada), 96

Operation Fortitude, 197
Operation Freya (German, 1975), 212
Operation Overlord, 208
Operation Rutter, 146–9, 154, 156
Orford Island, 10
oscilloscope 3339, 84
Oslo Report, 46
Osten, Captain Murray, 168–72, 173–7
Overlord, 208
Oxenden House, Leighton Buzzard, 45

Pankhurst, Chief Engineer Freddy, 40
Pankhurst, Sylvia, 40
paraboloids, 196
Parke, Keith, Air Officer Commanding 11 Fighter Group, 83, 92
Pearl Harbor, 97, 145, 213
Peenemunde, 195
Pegler, Westbrook, 211–12
Petit-Appeville, 177–8, 180, 183
Petty France, Westminster, 4
Philips, Colonel Eric, 96
Pierce, Corporal, 71
Pitt, Dr. Arnold, 99
Plendl, Dr. Hans, 30, 79
Polish Night Fighter Squadron 307, 118–20
Polish Squadron 317, 118, 126, 131
Pollard, P.E., 21

port orbit, 128
Portal, Sir Charles (later Lord), Chief of the RAF, 142
Pourville, 146, 156, 163, 167, 171–2, 177, 181, 185, 193–4, 196
Powell, Sergeant Bill, 120
PPI (Plan Position Indicator), 104, 114, 135, 137, 201
Pratley, Bill, 137
precision radar, 146
Prien, Lieutenant Günther, 48
Priest, Dr. Don, 8, 14, 23, 26, 80, 96, 103, 154–5, 171, 175, 213
Princess Beatrix, 160
proximity fuse, 99
pulse, 36, 38, 140
Pye 5 RF amplifier, 49

Queen's Own Regiment, Kingsbridge, 125
Queen's University, 98
Quilter, Sir Cuthbert, 1, 7, 13

Radio Research Establishment of the GPO (General Post Office), 10
RAF Fighter Groups
10 Group, 106, 108, 112–14, 117, 121–2, 135, 148
11 Group, 91, 93
12 Group, 93
60 Group, 45, 49, 65, 73, 80, 89, 106, 112, 148, 199
RAF Stations
Blenheim, 47
Coastal Command, Whitehall, 23
Exeter, 108
Signals and Radar, 142
Soar Mill, 106, 108
Uxbridge, 43

"railings", 26–7, 149
Raines, A.H., 3
Randall, J.T., 81–2
range cutting, 81–2
Rawnsley, Peter, 138
RDF (radio direction finding), 2, 16, 18, 80
RDX, 101
Rebecca transmitter, 207
Rechlin Research Establishment, 39
Research Enterprises, 99
Reynolds (1964), 212
Reynolds, Quentin, 86, 88, 160, 211–12
Rhubarb, 122, 126, 130–2, 157
Ribbentrop, Joachim von, German Foreign Minister, 39
River Scie, 168–9, 193
Robertson, General J.T., 196
Robertson, Terry, 212
Roosevelt, President Franklin D., 94
Rosehearty, Aberdeenshire, 53, 54–5, 63–5, 67, 70–1, 75–7, 102, 112, 132, 147, 155
Ross, Dr. John, 101
Rowe, A.P., 9, 10, 13
Royal Canadian Navy, 101
Royal Hellenic Air Force, 198
Royal Oak, 28, 48
Royal Radar Establishment, 51, 154

Sauster, Raymond, 212
SCR 584, 100
Sea Lion, 79–80, 83, 93–4
Seaforth Highlanders, 149
Seetakt, 141–3, 150, 199
Shoeburyness, 47
Signals Experimental Establishment, Woolwich, 21
"Silver", 156, 166, 177, 182

Slezak (Polish destroyer), 165
Smith, George, 14
"Smokey", 156, 174, 177–8
Soar Mill Farm, 107–8, 112, 114, 119
South Saskatchewan Regiment, 155–6, 160, 162, 164, 168, 171, 177, 185, 193
Spitfire, 95, 118, 123–4, 126, 128, 180
Stagg, Group Captain, 209
Stalin, Joseph, 53
Stavanger Air Base, 55–6
Stephenson, Sir William, 212
Stonehaven 51
Stones Farm, Waltham Abbey, 15
Suez Canal, 177–8
"Sunday Soviets", 13

Tait, Air Commodore Sir Victor, 142–4, 150–52, 154, 192–4, 197, 199, 204
Telecommunications Research Centre, 103
Telecommunications Research Establishment, 51
Telefunken Company, 33
"think tanks", 13
Thrussell, Private Leslie, 155, 167, 172, 215
Thurleston Hotel, 117
Tizard, Sir Henry, 9, 11, 12, 20, 95, 98, 101, 201
Tizard Harbour, 101
Tizard Mission, 94, 100
Toronto, University of, 98–100
Touch, G., 14
Townsend, Wing Commander Peter, 135, 137
TR 1082/1083, 65
transponder, 23
Trinity, 139–40, 194
Typhoons (aircraft), 157–8

U47, 48
U-boat war, 14, 96, 198
Udet, General Ernst, 31
UHF (Ultra-High Frequency), 18

V-2 rockets, 13
Ventral tanks, 126
VHF (Very High Frequency),
  18, 19, 69, 73, 148, 196
VHF 1132A receiver, 148
Vikings, 37

WAAF (Women's Auxiliary Air
  Force), 42, 57, 123, 136
Watson-Watt, Dr. Robert, 2, 8,
  10, 12–14, 16, 19, 26, 51, 80,
  83, 92, 97–8, 109, 213–14
Weber, General Willy, 214
Weedon, 10
Wehrmacht commanders, 31

Wellington (bomber), 137
"Wendall, Professor", 212
Wilkins, Arnie, 8, 10, 12, 213
Wimperis, Harry E., 9
"window", 141, 150, 194, 199,
  201
"wing" attacks, 92
Worth-Matravers, 51
Wright, Dr. George, 104
Würzburg, 32, 39, 46, 48, 84,
  141–3, 149–50, 195, 199, 202

"X" craft, 208–9
X-Gerat, 30–1
"X" raid, 51, 136

"Y" (listening) service, 143

Zeppelins, 33–5, 37–8, 88